KEIKO SPEAKS

KEIKO SPEAKS

Keiko's True Story
Based On His Communications
With Bonnie Norton

**By Bonnie Norton, Animal Communicator
and Keiko, the Orca Whale**
Who Starred in the Movie *Free Willy*

Animal Messenger Publishing

Keiko Speaks

PUBLISHED BY
Animal Messenger Publishing

Printed in the United States of America

ISBN.0-9763388-0-7-090000

COVER PHOTOS: Keiko - AP/Wide World Photos

&

Ocean Sunrise - "jefras"

* * *

COVER DESIGN: Mona Dinger & Bonnie Norton

TO ORDER BOOKS CONTACT

Animal Messenger Publishing

P.O. Box 275

Elgin, OR 97827

877-4-KEIKO-2 (877-453-4562)

www.keikospeaks.com

SPECIAL THANKS

This book is dedicated to Keiko, the incredible orca whale who starred in the movie, *Free Willy.* His giant heart touched millions of people both in person and on the big screen. Keiko believed his purpose was to teach humans about opening their hearts and loving all animals. I will always be grateful for my direct communications with such a magnificent being and for having the honor of being in his loving presence at the Oregon Coast Aquarium.

In the human world, I would especially like to thank Mary J. Getten, Teresa Wagner, Tyler Graham and Penelope Smith for their willingness to communicate with Keiko on a number of occasions, often on short notice. In addition, I greatly appreciate the cooperation of the many media sources who agreed to let me publish information from their work, including *The Associated Pre*ss and *The Oregonian* as major contributors.

I also thank Gary Collins for his constant assistance in documenting my work with Keiko on video; Clayton Beck who has and continues to keep current both of my web sites www.animalmessenger.com and www.keikospeaks.com; Steve Dicky who, during Keiko's five years abroad, kept his web site www.keikonews.com constantly updated with the latest information about Keiko; Keiko's long-time trainers Jeff Foster and Stephen Claussen, and the dozens of experts in the marine mammal community and governmental agencies involved with Keiko who, at various times, kindly gave me current information and their honest opinions about Keiko's situation.

In the print media, I thank Penelope Smith's *Species Link Quarterly Journal* (www.animaltalk.net); Merrit Clifton and Kim Bartlett of *Animal*

People newspaper; Judy Barnes and the *Sagebrush News*; and Ted Kramer, Dave Stave and Phil Bullock of the *Observer* in La Grande, Oregon for their ongoing support and writing of articles about Keiko and my work with him.

I also thank Portland's *AM Northwest*, Bend's *Good Morning Central Oregon*, and the many other people in television and radio who conducted interviews from 1998 through 2003. The willingness of people within all forms of the media to cover Keiko's story from the new perspective of Animal Communication was admirable–and without which I could not at all have spread Keiko's message.

I am deeply grateful to the following individuals who personally supported me and my work during the five years Keiko was the focus of my life: My mother, Beverly Swager; my father, John Holmdahl; Kathy Wahl, Vonnie Stefflre, Barbara Mark, Sally Field, Lisa Mason, Tony Anderson, Greg Dixon, Dawn Nelson, Michael Kloian, Sara Price, Mona Dinger, Irma Franklin, Dave Gillis, James Bradley, Misha Nosha, Michael Chocohlak, Ron Larvik, Chloe Larvik, Susan Meyers, and too many others to name.

And last but not least, Mim Lagoe for her patience, insight and willingness to edit this book, believing too, in the importance of maintaining the integrity of Keiko's story and his communications.

CONTENTS

PHOTOGRAPHS

ABOUT ANIMAL COMMUNICATION

In 1996 an Animal Communicator came to my riding stable and talked with several of my horses. I thought she was a one-in-a-million person who just happened to have this special gift. Later, I learned that we are all born with this ability, but most of us lose it at an early age. As soon as I heard that some Animal Communicators teach workshops so others can redevelop their natural ability, I attended every workshop I could get to on the west coast. When I realized I could help many more animals and people, I sold my barn and horses so I could become a full-time Animal Communicator.

To communicate with animals one must quiet the mind, let go of any preconceived thoughts or ideas, and allow the animal's response to come through. The communication is from the animal's perspective, and our intention is to receive the animal's own thoughts and feelings.

Communicators receive an animal's thoughts, concepts and emotions in a variety of ways. These include hearing, sensing, instant knowing, feeling what the animal feels, the animal sending mental images, etc. This varies with each person and sometimes with different animals. We simply honor however the communication takes place.

I receive most of my communications through hearing and then writing it down immediately. I usually hear one or two words at a time and have no idea where the thought or idea is going. Sometimes I receive the entire thought or sentence. Occasionally, the information comes in the other ways, and only rarely do I receive mental pictures. I always note in my communications if the information is received in a way other than hearing and writing. In whatever way it comes through, the person more or less interprets the animal's intentions and, in turn, the animal receives our intentions.

It is also possible to communicate long distance, somewhat like tuning a radio into the correct frequency. There is no physical connection between the two, but communication is possible.

If you are skeptical about our ability to communicate with animals you might be interested to know that the U.S. Government developed a research program at Stanford Research Institute called "Remote Viewing." The CIA, the U.S. Army and the Defense Intelligence Agency were involved in the program which included developing human telepathic skills to learn more about people, places and events at distant locations. They must have known there is some truth to this phenomenon or they would not have funded the program for twenty years at a cost of over $20 million.

As far as specifically communicating with animals, Rupert Sheldrake, Ph.D. has written a book, "Dogs That Know When Their Owners Are Coming Home—and Other Unexplained Powers of Animals". Using accepted scientific methods, he documents numerous situations where there is a connection between animals and humans for which he can find no explanation other than mental telepathy.

My dog, "White Socks" said it best during a communication with me: "We are not what most think or believe we are. We are creatures of love and joy, here to teach man of such things. Anyone who doubts you or what we are saying is just not ready to continue on their cycle of growth."

Animals are so much more aware than we have been led to believe. My hope is that Keiko's book will confirm this to the many people who already know it in their hearts, and that it will open the hearts of others to consider. May each and every person have permission to openly love and acknowledge animals for all they are. I believe that when people realize how intelligent and aware the animals are, they will be more loving, compassionate and respectful of all beings.

My hope is that my work will help create "reverence" for all animals. Webster's Dictionary defines reverence as "honor and respect mixed with love and awe."

* * *

"Until one has loved an animal,

part of one's soul remains unawakened."

~ Anatole France

INTRODUCTION

This is Keiko's true story based on my communication experiences with him between 1997 and 2003. Because of the volume of our conversations, some of Keiko's quotes were omitted, but none were altered. The communicator's dialogue was edited only for clarity and brevity. Except for my friend, Kevin, no names were changed to protect anyone. All information regarding Keiko's status was taken from newspaper articles, documentaries and/or television reports on him, or personal conversations with people concerned about or involved with Keiko.

During my work with Keiko, there were several times I asked two professional Animal Communicators to communicate with Keiko to verify the accuracy of the information I was receiving. Mary J. Getten and Teresa Wagner asked Keiko the same or similar questions, and only after I had completed my communication with Keiko did I ask for their results. Later, you will learn about Tyler Graham, who also contributed greatly to representing Keiko. All of our communications were taken either directly from their e-mails to me, or from our live or taped telephone conversations.

I always try to ask neutral, open-ended questions so as to not lead an animal to any specific response. Throughout the book you will notice that Keiko was asked similar questions a number of times. Because of his situation, it was important to know how he was feeling physically and emotionally; if he was eating enough to stay healthy; and how he was doing at different stages of his journey.

I also opened and closed most communications with the question, "Is there anything you would like to say?" This gave him a chance to say what was important to him before he considered my questions, and again before ending the conversation. I also silently opened each communication by asking him to communicate clearly and accurately because I wanted so much to represent him as truthfully as possible. There were also times I received two words to describe what he was telling me. Rather than decide which word was more accurate, those words are shown with an "and/or" slash mark.

In the same way, some of Keiko's responses were repeated over and over again—especially about his desire to be with people. I was faced with an editorial quandary as to how much of Keiko's plea to leave out. At the risk of being redundant, I decided to leave his responses as he communicated them to me and to others. I believe those who read this book will do so because they loved Keiko and want to know his true story, including his relentless cries to be with people.

Some of the questions asked during our group communications with Keiko might seem irrelevant during the conversation. This is because the questions were determined before the communications began, so we did not know what Keiko was experiencing at the time, or how he would respond to previous questions.

When our responses from Keiko differ, please keep in mind that each person interprets the animal's response individually. And Keiko, like any living being, can feel good or be happy on one day and not the next. As we know, experiences or circumstances can even affect one's mood several times during the same day. Also, because Iceland is seven hours ahead of the western United States (PST), and Norway is nine, Keiko may have indicated that he was not with people because it was the middle of the night.

The questions or comments to Keiko are printed in italics. Keiko's responses are in quotes. Comments from the communicator are in regular print with no italics or quotation marks.

Between 1998 and 2004, I wrote seven letters to the organizations responsible for making decisions for Keiko. I briefly tried to explain how Animal Communication works, and what Keiko had been saying. I made suggestions based on what Keiko wanted, and asked that they consider his communications and his behavior when making decisions for him. I also requested to meet with them personally with the hope they would see I am a relatively normal, sane person just trying to represent Keiko.

Four years later I received their first response, after which we wrote and spoke several times. Their general response was that despite Keiko's apparent desire to be with people, he was making great progress and they wanted to give him more time to choose to be free.

In an effort to write Keiko's entire story, initially I was going to include all correspondence between Keiko's organizations and myself. As the book took on a life of its own it became clear that, although this information was interesting, it distracted from Keiko's own message. I have included reports of what was happening to Keiko and what others said about releasing him, but other than occasional excerpts and references to letters I submitted, most correspondence between Keiko's organizations and myself has been omitted.

This is Keiko's book and I promised him I would do what I could to help him be heard. Keiko's "Message to the World" is contained in the last chapter and can be printed from my website www.keikospeaks.com.

KEIKO'S PAST

Keiko, which means "lucky one" in Japanese, was captured in 1979 off the coast of Iceland when he was about two years old. In 1982 he was sold to Marineland in Ontario, Canada, then sold again in 1985 to Reino Aventura, an amusement park in Mexico City. An estimated 25 million people (as many as 20,000 people a day) saw Keiko when he was in Mexico.

Keiko was chosen to star in the 1993 movie, *Free Willy,* about a captive whale who is set free by a young boy. At the end of the film, a telephone number was shown for anyone interested in saving the whales. Over 300,000 people called Dave Phillips of Earth Island Institute, and even more sent letters and donations insisting that Keiko be set free.

At the time Keiko was living in conditions far from his native cold water ocean at sea level. In Mexico he was at an altitude of 7,000 feet. His tiny pool was only 12 feet deep with 80-degree water, which also aggravated his virus-induced skin lesions.

When it was learned that Keiko was in poor health and possibly dying, $100,000 was raised to help move him to a healthier environment. Dave Phillips formed the Free Willy Keiko Foundation with the hopes of returning Keiko to health and eventually setting him free.

On January 7, 1996 over 1 billion people worldwide watched as Keiko was flown from Mexico City to the Oregon Coast Aquarium in Newport, Oregon. He arrived safely and was gently released into cool seawater for the first time in fourteen years. His new $7.3 million tank was five times larger than his tank in Mexico.

This is where my story with Keiko begins.

The Day I Met Keiko—
My Life Changed Forever

August 16, 1997

This was a day I will remember for the rest of my life. It all began on a warm summer afternoon during a drive to the coast with Kevin. Not having been to the Oregon coast before, I asked him what there was to do there. Knowing my love for (or obsession with) animals, Kevin said, "Well, if you want, we could stop by and see Keiko." That year I had been attending workshops to learn how to communicate with animals, so my first thought was, "Wow, I wonder if I can communicate with a whale?"

I didn't even know where Newport, Oregon was and I was on my way to see Keiko, the famous orca whale. At that time, I hadn't seen the movie, but I remember watching the television news when Keiko was flown from Mexico to the Oregon Coast Aquarium. I cried as I watched the staff standing in his new tank, helping him out of the sling that moments before had lifted him over the aquarium wall. Keiko was very quiet until the end when he wiggled around a little, then swam out of the sling into the cool water of his new home. It was heartwarming to see people do so much to help an animal.

Now, a year and a half later, Keiko was much healthier. He had gained weight and his skin lesions had healed significantly. Reports were that the Free Willy Keiko Foundation, the organization responsible for Keiko, was beginning to make plans to return him to the wild. I thought their intentions were quite honorable.

As Kevin and I entered the aquarium, we were warned that Keiko was not always visible in his tank, and that no refunds would be given if we could not see him. Although his tank was relatively small, if he was far enough away from the viewing windows, he could not always be seen through the murky water. How disappointing it would be to be right there and not be able to see him.

As we walked towards Keiko's exhibit, some people walking away were commenting that they had waited for a long time and still did not get to see him. Even though I wasn't sure what to expect, magic was already beginning to happen. I was becoming more excited with anticipation, hoping that I would be able to see Keiko.

We slowly followed the crowd into what felt like a cool, underground cave. Keiko's exhibit had three large viewing windows, one in the center and one angled on each side. It was a little dark and crowded, but all other awareness disappeared when I spotted Keiko. There he was—slowly moving in front of the window on the right. There were three staff members in wet suits swimming in the tank. Now that I think about it, they must have just gotten in the water because they were where the spectators could easily see them and Keiko.

We stood there in awe at Keiko's magnificence. His beautiful black and white body was like nothing I had ever seen before. I noticed his bent dorsal fin, which had become his trademark, and three black dots under his chin. Now, as I write this, I realize that the overwhelming feeling I had when I first saw Keiko was much more than just his striking body. There was something about seeing him, just being in his presence that touched me on a much deeper level. I didn't realize it then, but Keiko touched my heart in a profound way. That day I fell in love with a whale and, little did I know, Keiko would change my life forever.

Kevin watched for a short time, then left. We agreed he would come back in about an hour. I spotted a place on the railing around the back of the exhibit where I could set down my notebook and, if I could communicate with Keiko, write everything down.

As I started making my way through Keiko's tightly packed fans, I mentally said, *Hello Keiko. My name is Bonnie and I am here to see if you will talk to me. I want to tell the world what the animals have to say.* I no sooner finished the thought when, to my surprise, Keiko replied, "Start writing. I'm ready!"

Keiko interacts with some of his many admirers at the Oregon Coast Aquarium. AP/Wide World Photos

I quickly flipped through my notebook to a blank page and began writing every word.

"We would like man to realize what he is doing to the world. That many changes have taken place over years and there are more things to come. Man must realize that he has a tremendous effect on the future of our planet."

Then I heard and wrote one slow word at a time.

"Where…are…you?"

First, I was surprised by the sudden change in topic; then even more surprised by Keiko's awareness. He was asking me where I was! It was almost eerie. I left my notebook and made my way towards the window, almost as if, somehow, I might miss him. As I got closer, I stood in the center of the glass viewing window on the left side. There were fewer people by this window and, being fairly tall, somehow I thought he would know me. Keiko swam past me in a counterclockwise direction following the last diver who then got out of the water.

I mentally said to Keiko, *You just swam past me, I'm in the center of the first window.* Then I repeated, *You just swam past me, I'm in the center of the first window.* As he completed his circle to the left, he turned and swam straight towards me! I'd swear he was looking me right in the eyes!

He turned left just in front of the glass. I stood there for a moment, just trembling. The physical and emotional sensations were indescribable. A young boy standing with his father next to me excitedly said, "Daddy, daddy, he swam right at us! He swam right at us!"

My heart was still pounding as I slowly walked back to where I had left my notebook. Overwhelmed by my experience, I said to myself (not doubting what had just happened, but as one would say in pure amazement), "I don't believe this." To my surprise, Keiko immediately responded to my own mental thought.
"It is your connectedness with the universe and the animals that allows this. Don't doubt yourself anymore. This should clarify it for you!"

It took me a minute to collect myself and start writing again. I told Keiko that I hoped to write a book to let people know what animals have to say. I asked, *What else would you like to say to me or the world?*
"We would like man to understand that we all want to live in peace. To be free from fear and pressure to survive. We want all animals and man to love alike. There will always be differences but those can be worked through in a peaceful way—without violence and war. Understanding and compassion are the answer to peace. If man only knew how much he is influencing everything on this earth, and everything he does will leave an impact on all other living beings."

Keiko's thoughts flowed easily like a wise, old man. *How can we or you or I help others to understand?*
"You are a start and others before and after will help shift the world in a better direction. Your book is a wonderful idea and many will listen and learn and grow from what they hear from us. We are all here with a/our specific purpose to learn and grow and be who we truly are; to learn to love and forgive and enjoy our life on this earth. How many people do you know who truly live their lives that way?"

Probably some, not a lot.
"See what I mean? People are stumbling through their life with no awareness of their purpose or the big picture. You will be the connection

for them—to help others find their way in life...You can ask other questions."

Feeling woozy, I said, *Wait a minute Keiko, I need to regroup. This is all quite overwhelming. I need some air.*

As I walked outside, Keiko continued.
"Good for you. You know how to take care of yourself—something else humans need to learn more of how to do."

I took a few deep breaths of the warm, fresh air and then returned to my post inside. *Keiko, I'm back inside. Is there more you would like the world to know? What is the most important thing you would like the people of the world to know?*
"Learn to live in peace and take care of our world. It does not exist just for man alone. Many others depend on the earth and its survival for their existence. Man believes he is the special one and all others are less significant. Without the rest of us, man is nothing—he will cease to exist."

As a way to validate that I was truly communicating with Keiko, I asked, *Is there anything else, or something specific you can tell me about yourself?* I was hoping he would tell me something I would have no way of knowing, or something only those working closely with him might know. I was very surprised by his response.
"I am quite content for the moment. I am feeling much better and healthier than before. I like the attention and effect I have on all the people that come to see me, especially the younger ones and the children. They will remember seeing me for a long time. If they set me free because they believe they are helping me, they will be doing a disservice to many, many people that will be influenced. I am happy enough here to stay and have an impact on the many people who will visit me. I am willing to stay, versus being set free for me alone. I need to impact others by my captivity much more than I need to be set free. If they insist on setting me free, I will continue to do things and/or behave in a way to discourage them from letting me go, for my purpose is best accomplished by being here. Please let them know this. They believe they are doing what I want, but it is what they want for me. They mean well, so please help them to understand me."

For a moment, I considered it might help to have another Animal Communicator validate his request. Again, Keiko responded to my thought.

"Yes, bring in other Animal Communicators so they will believe you and me. Love to the World from Keiko."

I walked outside, sat down with Kevin and just cried. I told him, "You'll never believe what just happened!" Without a doubt, this was the most profound communication I have ever had with an animal. At that time, I wasn't sure what I was going to do with the information from Keiko. I had heard talk about setting him free, but for some reason during my visit it didn't seem like it was really going to happen. And if it was, not any time soon.

To Free or Not To Free Keiko

Fall 1997

After my profound communication with Keiko on August 16, 1997, I felt compelled to ask just about everyone I talked with how they felt about returning Keiko to the wild. To my surprise, almost everyone responded that they did not think Keiko should be set free. Some commented that because he had been in captivity since he was two years old, that it would be like taking their dog out in the country, turning him loose and expecting him to survive. Many said that "Keiko just loves people and seems happy at the Oregon Coast Aquarium." I would then tell them what Keiko said he wanted. Again, total agreement.

Their comments continually verified why Keiko said his purpose in life was to be with people. Keiko knew how those who connected with him and saw him in his small tank immediately knew this is not the way he or any wild whale should live. Keiko believed his purpose was to teach people about opening their hearts and caring how whales and other animals, should (or more obviously) should not, live. Many people told me that they really didn't care before, but after seeing Keiko they became dead set against capturing or killing whales of any kind.

Those who saw Keiko talked about how much he touched them on a deeper level, and that they "will never forget seeing him." Even people who had not visited Keiko said they did not think he should be set free. It was as though, intuitively, people just "knew" in their hearts what Keiko really wanted.

Over 2 million people visited Keiko at the Oregon Coast Aquarium during the 19 months he considered it his "home." Many times there were 400-500 people with him all day long.

I began reading every article I could find about him. I would call each specialist interviewed in the articles, and ask what he/she thought of releasing Keiko. Each would give me their opinion and then the name of someone else to contact. One call led to another and I ended up talking with dozens of people all over the country. I spoke with marine mammal specialists in the U.S. Navy, the U.S. Marine Mammal Commission, National Marine Fisheries Services; as well as presidents of aquariums, Keiko's previous veterinarian, and trainers who had personally worked with him.

Most said that releasing Keiko would be impossible because the most important criteria for successful release of marine mammals is "minimal human contact." They explained that the longer an animal has been with people, the more likely it is to follow boats and jet skis. Often these animals have had positive experiences with people and come back to them for food and attention. They become labeled "nuisance animals" and put themselves at risk of being injured or killed by boats. The overall consensus was that Keiko's release was not likely to succeed, and that it could be very risky for him. In addition, if Keiko's release was not a success, future attempts to release captive marine mammals may be less likely.

As time went by, the few people I did speak with who originally supported freeing him, said they only wanted it to go as far as was reasonably safe for Keiko, and if he had a good chance of surviving. After it became clear that Keiko did not want to be free, they quickly changed their support to his obvious need and desire for care and attention from people. Having talked with hundreds and hundreds of people between 1997 and 2003, I can honestly count on my fingers (I don't even need to use my toes), the number of people who truly believed the attempt to free Keiko should be continued.

Some people believe all animals should be free, not just what we call "wild" animals. One person even told me it was better for Keiko to be free for one week or one month, even if he died, than to live another day in captivity.

Keiko should never have been captured in the first place, but he was. And, we are responsible for the lifetime happiness and well-being of any animal we take into our custody.

In general, I do not believe in keeping wild animals in captivity. However, by communicating with them directly I have learned that some animals want to be with people, and they come to us to teach us and heal us. Maybe the goal should be that someday the only animals in captivity will be those rescued or that "come to us" for help and rehabilitation.

Because so many people knew and loved Keiko, I believe he will be one of the biggest teachers of all. He has shown the world how important it is to listen to the animals and honor them as individuals.

The following article by Katy Muldoon, *The Oregonian*, Sunday, November 2, 1997 described the controversy. It was titled "Lack of data, questions of ethics muddy waters on Keiko's future,"

> …Around the nation, marine mammal experts—who seem largely loath to publicly enter the Keiko fray—voice intense concern about releasing the world's most famous whale. And they cannot agree on whether a bay or sea pen is the best next step for Keiko the controversial, as the whale's owners propose.

> Scientists caution the release has the potential to spread disease among wild whale populations and to compromise the strength of the species through unwanted DNA exchanges.

> They fear that an animal that has spent far more years with humans than he has with whales is a poor candidate for such an experiment—an animal with little chance to learn adequate survival skills or successfully fend for himself.

> …Marine mammal specialists have learned in the intervening years that rescue, rehabilitation and reintroduction need to be swift, in part so the animals retain their inborn survival skills.

> With the exception of Hollywood's "Free Willy" film version, no one ever has released a long-captive whale.

> The effort to mimic the movie by rescuing and possibly freeing Keiko has captured Americans' hearts and persuaded them to open their pocketbooks. But many who work with marine mammals believe those people have been misled—that it is a foolhardy cause, an impossible task that would not hold up to scientific scrutiny. "It would be like sending your wheelchair-bound 87-year-old grandpa out on his own," said Tom LaPuzza of the U.S. Navy Marine Mammal Program in San Diego. Most scientists believe that if the animal were let go, it would be dead in two days.

…Most of what is known about reintroducing marine mammals to the wild comes from work at facilities such as The Marine Mammal Center a few miles north of San Francisco.…

…Scientists there and elsewhere have learned that keys to successful rehabilitation and release of wild marine mammals include such elements as minimal contact with humans plus full ability to feed themselves, socialize and act like the aggressive predators they are. Those rules may apply to captive animals like Keiko, too. In his current condition, the whale fails the releasability test on every count.

…The foundation has not wavered in its much-publicized dream to free the whale. However, Nolan Harvey, the marine mammalogist who oversees Keiko's daily care for the Free Willy Keiko Foundation, and others have said all along that they intend to do what they believe is best for Keiko as they observe his progress—even if that means keeping him in captivity for the rest of his life.

Despite mounting criticism from the scientific community—plus the foundation's recent dispute with the aquarium over contracts, money and the whale's health—he and his staff remain steadfast in their mission:

"The idea," he said, "is to see exactly how far Keiko can go"… Somebody has to step forward and lay the groundwork for scientific protocol. Unless we do this, we're never going to know if captive whales can go free."

…Even some who have pushed for Keiko's cause question whether he was the best candidate for such a project…

KEIKO AT THE OREGON COAST AQUARIUM

1998

I began hearing more about Keiko and the Free Willy Keiko Foundation's plans to move him and then set him free. I knew I couldn't stop them, but continued to tell people what Keiko wanted. I found almost everyone was interested in Keiko.

In January 1998, I was at my local Humane Society adopting one of my dogs, "White Socks." Of course, I began talking with the manager about what Keiko had said. She responded with what had become an almost universal response, "I don't know why they are moving him anyway. He seems perfectly happy where he is." If I had a dollar for every time I've heard that, I'd be rich!

Anyway, it was on this day that I had two unusual communication experiences. I was in a hurry and quickly walked from my bedroom past my lovebird, "Phil," into another room. Without any solicitation from me, Keiko said, "Thank you for spreading my words. I appreciate it. Other communicators are not saying enough…I would like to talk with you again."

I told Keiko that I wanted to see him again too, and that I would visit him as soon as I could. As I rushed back past Phil, he said, "You're going to see Keiko?" I stopped dead in my tracks, turned around and said, *You know of him?* Phil said, "Of course, everyone knows Keiko."

February 1998

Craig McCaw, a Seattle telecommunications billionaire, donated the majority of the money to free Keiko. He also became chairman of the Free Willy Keiko Foundation board of directors.

The following article titled "McCaw's world view extends to Keiko," by Steve Woodward and Katy Muldoon, *The Oregonian,* Sunday, February 15, 1998, featured Craig McCaw:

> "The high point of my entire life spiritually is spending time with Keiko in the pool. It's almost like dealing with an extraterrestrial, having a relationship with an intelligent being on a very personal level."

> …He believes that mental telepathy will become an everyday reality.

In February, Jean-Michel Cousteau, son of internationally known French ocean explorer Jacques Cousteau, also joined the Free Willy Keiko Foundation board of directors.

March 26, 1998
It had been a long time since I had communicated with Keiko, so I wanted him to know I was on my way to see him.

Keiko, this is Bonnie. Do you remember me from earlier?
"Of course I do. I remember/know many, many things."

I am coming to see you and want to check with you first.
"I already knew you were coming. You told me before. I am looking forward to our interaction. I have lots to tell you and show you too."

Do you still want to stay at the aquarium versus being set free?
"Yes, I still feel as I told you before."

Do you realize that they are trying very hard to help set you free?
"Yes, I realize that but it won't happen."

Why?
"Because it can't happen."

Why?
"I won't allow it. I belong here. This is my home.*"*

That's why I'm coming to visit you again. We need to talk more of how to keep you from being moved and set free.

I was pretty tired from traveling so I told Keiko, *I feel I need to continue*

our conversation when I am there with you the day after tomorrow. Is that okay?
"Yes, I am looking forward to seeing you again. I will give thought to our dilemma and have answers for you."

Thank you, Keiko.

March 28, 1998

My heart sank as I entered Keiko's exhibit—no Keiko! It was one of those days that Keiko's pool was murky and it was impossible to see him if he was very far from the viewing windows. I tried not to be too disappointed knowing I would be there for the next couple of days, and surely would be able to see him again later.

I set my notebook in the same place on the railing, looked hard into the cloudy water and then began communicating with Keiko.

Hi, Keiko. This is Bonnie. I have come to talk with you. As happened before, Keiko responded to my previous thought.
"You don't need to see me to speak to me."

Yes, I know, although it would be a great sign and I'd love to see you. I'm having trouble trying to help you with your request that you do not want to be set free. Do you still feel the same? My body began trembling.
"It's okay, you are doing fine."

You know, I cried, watching your whole movie. At that time, Keiko swam into view for all to see. *Thank you for coming in! You are so beautiful!*
"Did you hear all the gasps of amazement when I appeared?"

Yes, I did.
"That is why I want to remain here. I love my work here. See how touched everyone is? They will remember me for a long time. They will learn what not to do with us. If they let me go I will not have the influence on these people."

I understand, but do you realize that they are trying very hard to set you free? In fact, they are making plans to bring you to a bigger pen in an ocean bay. They hope you will show them that you can survive in the ocean on your own. I will check that out so I can give you accurate information. If it is true, would you want to stay there without people seeing you?

"How depressing not to have my fans/friends that come to see me. I would not want to move if that is the case."

Do you still want to stay here?
"Yes, I love it here."

How do we keep your people, those who ultimately control your destiny, from moving you from here?
"Don't worry, I know what to do. I have all-knowing about these things. I am acting rather stupid about my food. Watch this, I'll swim right by my fish dinner. How do they expect me to survive? Not only eating, but all the other elements as well. I am fine where I am. Why can't they understand that?"

Probably because everyone thinks you would rather be free.
"It's a human dream—not mine. I would rather be here with my human friends."

Do you want me to print/publish what you have told me now and before?
"Yes. I do want everyone to know what I have said."

How do I do that so that your people support it?
"Go to the Foundation—that will help them admit/accept that I am better off here. They know it in their hearts, but they keep trying anyway. Help make their decision easier. It will also take the pressure off me to be and do more. I'd like to just enjoy my life here with people and not have to learn more tricks. It's okay for them to study me and learn from me, but I'm tired of being challenged to be more/better than I am here. I can do whatever they want, but I'd rather just be here enjoying the people who love me. I have worked hard before. Isn't that enough?

Yes, of course you have. Keiko, I will check at the Foundation office and try to find out more about their plans for you. I will visit you again tomorrow, my friend.

As I left Keiko's exhibit, I wrote down what was written on a marker board by the exit:

> Independent experts will analyze the research to help the Free Willy Keiko Foundation determine if and when Keiko is ready for release. Whatever is best for Keiko is what will be done.
>
> ~ *Free Willy Keiko Foundation*

March 29, 1998
That morning I met outside with one of Keiko's trainers. I told him that I had been communicating with Keiko and asked if he was interested in what Keiko had to say. He said something to the effect that they were all scientific people there and didn't believe humans could communicate with animals. He suggested I contact someone at the Free Willy Keiko Foundation office.

When I entered Keiko's exhibit he was quietly floating by the center window. *Hello Keiko. This is Bonnie. What would you like us to know?*
"I am doing well. I like my home here. I have thought about our conversation and would like everyone to know that I do not want to be set free. I am willing to go wherever I can influence other people. That is my purpose in life. I will continue to behave in ways less than I am capable, because I do not want to be set free. My goal is to let the people of the world learn by my captivity. Learn what not to do to us. Learn that we are born free to stay free—never to be caught and made fools of. We are far more than man realizes."

Keiko, the information I have at this time is that the pen for you will be accessible to some people, but not nearly as many as here. If that is the case, would you rather stay here or is it okay for you to move there?
"I want to be with the people—whatever that means. Make my existence worthwhile. I have dedicated my life to teaching others about us and the ways of the world. We can only do so much; after that, it is up to mankind."

Is there anything else you would like to say?
"I want everyone to know that I thank them for all they have done for me. I am happy here and wish they all would be as happy with me here as I am. Conflict is not good for any of us. Please learn to live in peace and love with all."

That afternoon I visited the Free Willy Keiko Foundation's office adjoining the Oregon Coast Aquarium. There was a small model of Keiko's bay pen representing the one that was currently under construction. I was surprised to see how fast everything was moving. It became clear that Keiko would soon be heading somewhere in the Atlantic Ocean, possibly his native waters of Iceland. There would be a "viewing" area from atop the rocky cliffs surrounding the bay. People would be so far away, I couldn't imagine anyone making a trip to Iceland to spot Keiko through a telescope in his bay pen.

March 30, 1998
I arrived early that morning and was surprised there were so few visitors. I remember Keiko "hanging" vertically next to one of the viewing windows, and that he didn't move much during most of our conversation.

Hi, Keiko. I'm back. This is Bonnie. I have been doing some research on their plans for you, and I'd like your input. They are trying to find a bay in the Atlantic Ocean near where you were born. They believe this is the best place for you if you are to be set free. They are designing a bigger netted area where you could swim in the bay and come back to your own area. Then, if it appears safe and you want to, you will be free to live in the ocean. They feel your freedom and ability to be with other orcas is important to you. What do you feel about your future destiny? What do you want?
"I want them to know that I believe my purpose is to be with the people. They are really my family, I know little of my own."

Also, in this new area, you would not see as many people as you do now, only trainers, caretakers and researchers.
"I wish they would all leave me alone. I am happy here the way things are. I have had enough trauma in my life, don't you think?"

Yes, I do, Keiko. So what would you like and how do I help you accomplish that?
"I would like you to represent me. They all listen but do not hear me. They want me to be something that I'm not. I am not a wild whale; I am Keiko."

Somehow, I will get your message to the world. Right now I'm trying to figure out how to do that and convince your people that you do not want to be set free. Both will be difficult, especially the part about keeping you here.

At that point Jack Smith, a photographer with *The Associated Press*, entered and began taking pictures of Keiko. I approached him by the viewing window and introduced myself. He said he had been "on top" (above us and outside at water level) with Keiko's trainer. Keiko was supposed to be showing them how well he was progressing for his life in the wild. But, Jack added, "All I know is he wasn't doing anything they wanted, so they put him in 'time out,'" as he nodded at Keiko hanging next to the window.

Then Jack said, "Putting him in 'time out' doesn't really work either—Keiko just comes down here with the kids. He loves the kids—just line them up and he'll come over to see them."

I asked Jack, "Any chance Keiko is trying to tell everyone what he really wants?"

Like so many others, Jack said something like, "Personally, I think he's happy where he is."

A volunteer who made occasional announcements over the PA system had joined our conversation. She said when they fed Keiko a live fish the other day, he opened his mouth and let it swim out.

She also mentioned that Keiko had a television he could watch through the underwater office window on the other side of the pool. She said Keiko was not interested in watching videos of other whales, but enjoyed watching high-action movies like *Star Wars* and *Independence Day*, along with westerns, pro-wrestling and cartoons.

They both left and I finished my conversation with Keiko. *So, Keiko, the reporters are here to show how you're progressing, and you are not cooperating—how brilliant of you!*
"I'm doing pretty good, don't you think?"

Definitely. What specifically do you want me to do?
"Keep on your path. You are doing what you know needs to be done."

At this point Keiko swam upside down straight towards me. *What do you want me to do?*
"Why do you keep needing such proof? You do your part and I'll do mine. Together we can make a difference."

He began pushing a blue ball with his nose. *Keiko, I need to go home now. Is there anything else?*
"I have to do something right or they'll all be upset with me. Thank you for your concern and help in finding my dream."

On my way out, I stopped in the aquarium gift shop and purchased a 3-foot stuffed "Keiko" whale. He is beautiful—black-and-white velvet complete with bent dorsal fin and three black spots under his chin.

Keiko playing with a live crab before pushing it out with his tongue.
AP/Wide World Photos

April 1, 1998
Two days later, April 1, 1998, *The Associated Press* released an article by Brad Cain titled, "Movie's Killer Whale Not a Hunter":

> The romantic notion of "Free Willy" star Keiko swimming back to the sea just like in the movies has run up against a hard reality: The killer whale doesn't have much of a killer instinct.

> Keiko is so lousy at catching fish that even some of his most optimistic handlers acknowledge that the whale who has lived nearly his entire life in a pool may never be able to survive in the wild.

> He is getting lessons from humans in catching fish…

> …But for the most part, the black-and-white beast watched meal after meal flutter by.

> …"I don't think it's fair and humane to the animal to try this

operation just to make a few people happy," said Brad Andrews, vice president of zoological operations at Sea World in Orlando, Fla....

Andrews said his team of marine biologists has rehabilitated and released whales, usually ones that have been found beached or have other problems, such as malnutrition.

"But those animals are with us a very short time and not long enough to be imprinted by humans," he said. "Keiko has been imprinted a long time."

Though disappointed by the session, Keiko's trainers said it's still remarkable that a captive whale that's been fed dead fish practically its entire life now is able to catch and eat at least some live fish.

They said on better days, Keiko catches up to 15 live fish, or about half his daily diet.

"To my knowledge, no other project has attempted to teach an adult killer whale how to hunt," said Diane Hammond, spokeswoman for the foundation. "We're breaking new ground here every day."

She said the foundation believes Keiko would still be much happier in an ocean pen, even if he had to spend the rest of his days there.

"He will remain under our care as long as he needs it," she said. "If that means forever, so be it."

May 3, 1998

This communication with Keiko started unlike any other. I woke up feeling the physical sensations of floating, turning, undulating and spinning in water. I felt somewhat disoriented or confused, but definitely felt the connection was with Keiko. I grabbed my pen and paper and asked Keiko, *Is that the way you feel?*

"I am scared. I know much is happening and things are changing. I'm afraid to leave here. I don't want to go. Please help me."

Should I contact the aquarium or the Foundation first?

"I don't know about those things. Why are they trying to do this to me? Why do they want to take me from my home? I love it here. I am happy here. I love the people that come to see me. Why? Why? Why?"

They think they are doing what you want.
"Please tell them it is not what I want. I want to stay here. This is my home."

I'm going to work for you now, Keiko. I will do whatever I can to help you stay at the aquarium. Please keep in contact with me about whatever is going on for you. By the end of our conversation, I felt a playfulness about the movement, as if Keiko was playing and moving with joy and excitement.
"Thank you, thank you, thank you for helping me! I love you!"

I love you too, Keiko.

May 16, 1998
How are you, Keiko?
"I am doing better because I know you will be helping me. I can see a time that we will become friends and meet in a physical sense."

What about their plans to move you?
"They are working hard for that, but I am resisting by showing them that I cannot go along with their plans. They are so eager for me yet I have no desire. They are frustrated at their (or my) unsuccess. I wish they weren't so concerned with the outcome. I can't believe there is so much pressure to move me. I have it all here—this is my home! I can't understand their obsession with moving me."

What would you like me to do?
"Can you please let them know how I feel—that I love it here and want to stay here. I love my work here and would be very sad to leave. Why can't humans get along and see the value in working together? Where are things with you and me?"

Not too far I'm afraid.
"I need to do my work here with the people for as long as I can. That is much more important than my time of physical survival. I do not care about that. I only care about what I am able to do while I am here with the people."

Is there anything else you would like to say?

"I am grateful for anything you can do for me and the people. I can only hope that they will really listen to me through my behavior. They know me so well they must understand what I am telling them. If only they will listen! Thank you for whatever you can do to help them understand me. I love you for your help and them for their care and devotion. You are all wonderful people to dedicate so much of your life for me. Please thank them, too."

August 1, 1998

For some reason I have notes (rather than Keiko's exact words) that were written down shortly after a communication with him on August 1, 1998. Since Keiko had said he would "continue to do things and/or behave in a way to discourage them from letting me go," I asked him about continuing to refuse to eat live fish as a way to show them he couldn't be set free.

Keiko told me that he didn't want to kill the fish. Knowing how odd that sounds coming from a whale, I didn't tell anyone what he had said.

August 11, 1998

On August 11, 1998 there was a great article detailing Keiko's planned move. It was titled, "Keiko will take to skies Sept. 9 for big trip," by Katy Muldoon, *The Oregonian*:

> As they said more than 2-1/2 years ago in Mexico: "Adios, Keiko."

> The well-traveled whale is scheduled to leave Newport about 6 p.m. Sept. 9 and arrive in Vestmannaeyjar, Iceland, about 9:30 a.m. the next day. The flight will take about 8 hours.

> The Free Willy Keiko Foundation on Monday announced the whale's elaborate travel plans. Enough to give a travel agent a headache, they involve flatbed trucks, an enormous plane, a barge and several cranes.

> The foundation, which owns the whale, will cover the cost of the flight—approximately $200,000.…

> …Late in the afternoon, keepers will ask Keiko to swim into the same heavy-duty fabric sling that lifted the 21-foot-long killer whale from his pool at Reino Aventure, the Mexico City amusement park he left for the larger, colder pool in Newport. Keiko is 16 inches longer than when he arrived in Oregon and his head will

poke out one end and his flukes the other. The sling has holes in the sides to allow the whale to move his pectoral flippers.

The sling will be hooked to a crane capable of lifting 40 tons; Keiko weighs about 5 tons, or 10,000 pounds.

It will lift him from the water, over the walls of the tank and into a new, custom-made, 28-foot-long steel and Fiberglas container shaped — gulp! — like a coffin. Keepers prefer to call the container a "cradle," and it may feel more like that to Keiko. The whale will be suspended in cool water and ice that reaches about the level of his mouth; on the plane, a water tank will allow keepers to drain off old water and add new.

The boxed-up whale, aboard a United Parcel Service flatbed truck, will travel at 10 to 15 mph along Southeast Ferry Slip Road and turn south on U.S. 101 for the 3-1/2 mile trip to the municipal airport; the highway will be closed in both directions during the transport, and it likely will be lined with hundreds of Keiko fans, as it was on the blustery January day the whale arrived in Newport. Except for the news media, onlookers will not be allowed at the airport.

There, Keiko will be loaded into the belly of the cargo plane.

…The entourage flying with the former film star whale will include several handlers, four veterinarians, including Lanny Cornell, Keiko's longtime veterinarian, and members of the Foundation's staff and board of directors.

They should plan to bundle up. In order to keep Keiko comfortable, the crew will keep the air temperature in the plane at 52 degrees, which should feel to the whale like a typical Newport morning.

In flight, handlers will rub the whale and scratch his skin, which feels hard and smooth, like a fully inflated inner tube. They will monitor his heart rate and respiration, though there is little they can do if Keiko gets into medical trouble during transport.

The whale, however, is an old hand at traveling. He was captured in Icelandic waters at about age 2. Three years later, Marineland in Canada bought the whale, and he took his first flight. Later, the whale moved to Mexico City and then to Newport.

When Keiko arrives in Iceland, another UPS truck will meet his plane and drive the whale a couple of miles to the harbor. From the dock, a crane will load the whale onto a barge, which will travel a short distance outside the breakwater and into Klettsvik, the sheltered bay where the pen is moored.

A crane on the barge will lift the whale into his pen, and when the sling is released, Keiko will swim in North Atlantic waters for the first time in 17 or 18 years.

…Originally, the foundation intended through a lease and operating agreement with the aquarium to continue rehabilitating marine mammals in Keiko's tank long after he moved on. However, the two organizations never fully recovered from a bitter dispute last summer and fall, when they argued over money, power and the whale's health.

August 22, 1998
On August 22, 1998, I sent a letter to every member on the Free Willy Keiko Foundation Board of Directors as well as the Oregon Coast Aquarium, the Humane Society of the United States, Warner Brothers, UPS, and Reino Adventura, all of whom were involved in moving Keiko from Mexico to Oregon.

I thanked them for helping Keiko, explained that I had been communicating with him, attached excerpts from our conversations regarding where he wanted to live and why, and asked them to contact me. At the time I felt I had made the offer to open the lines of communication and wanted to let them know what Keiko had said. No one from the Free Willy Keiko Foundation or any of the other organizations contacted me about Keiko's communications or my offer to work with them.

During the last 2 weeks before my final visit with Keiko, I was pretty tired and stressed out from gathering information, writing letters, making phone calls, etc. There were a number of days I never even changed out of my nightshirt—I just got out of bed, went into my office, worked for him all day, then fell into bed late at night. During the last two weeks, I even told Kevin that I didn't have enough time to see him (I was too preoccupied emotionally anyway). All told, I spent over 6 weeks trying to let people know what Keiko said he wanted.

At times, I thought if I could just reach enough people or the right person, maybe someone would listen to what Keiko wanted. Then again, realistically, I knew there was no way I could stop them from moving him. I remember waking up early one morning about a week before Keiko was to leave for Iceland. I seriously asked myself, "Why am I doing this? I can't stop it!" The answer came to me immediately—actually two reasons:

First, the world will then know how Keiko truly feels and, perhaps for the first time, hear that it is possible for people and animals to communicate with each other. When people hear what Keiko has said, and then watch him validate it through his behavior, even skeptics will remember that he actually did tell someone he did not want to be set free. Because of the way the media followed Keiko, and because so many people knew and loved him, I realized Keiko could be a way to help many other animals by increasing human awareness that we can communicate with them.

The second reason was that I promised Keiko I would do what I could to get his "Message To The World" to as many people as possible. At that time most of his message was the worldly advice he offered in our first few conversations about humanity, peace and our planet. The last paragraph is a combination of communications in 1997 and 1998 about his purpose in life and how he felt about leaving the Oregon Coast Aquarium. *Keiko's Message To The World* can be found in the last chapter of this book and can be printed from my website www.keikospeaks.com.

At the time, I had no idea what the future held for Keiko or that we would ultimately engage in so many conversations. I now believe *Keiko's Message To The World* contains all of his wisdom shared during the years we communicated as represented in this book.

In the 2 weeks before Keiko left, I was interviewed on five Oregon television news programs and a number of radio shows. It's not something I ever wanted to do, but knew it was something I had to do. At the time, it was the best way to speak out for Keiko.

After a television interview at home in Bend, Oregon, a reporter said to let him know if I was going to see Keiko at the aquarium before he left. Since I was, he said if I sent out a press release the television stations would probably show up.

This gave me an idea and, after talking with Vonnie, my long-time friend with great insight, I decided to do a "rally" at the Oregon Coast Aquarium—again to let people know that Keiko wanted to stay and be with people.

Although the television and radio media were wonderful to me, and practically every individual I spoke with supported my work for Keiko, I couldn't figure out any way to get Keiko's decision-makers to listen to him. I had hoped bringing it to the attention of the public might spark some reaction. I received lots of support, but no one really willing to do anything.

August 26, 1998
It was about 2 weeks before Keiko was to leave the Oregon Coast Aquarium. This communication was more of a monologue; Keiko had a lot to say and didn't need much prompting from me.

"I would cry for the people and myself. There is a huge injustice being done here. I have been doing fine here. I love this place and prefer to stay. If they move me, there will be a great disservice to many including myself."

"I am able to handle whatever happens to me. I can be resilient at times if there is a purpose. I feel sad to leave my home and hope that something can be done for me and the world. I feel inadequate for this move. There is no rational reason for it, just because some think they know me better than I know myself. I believe my life has had purpose; so will my death. I am ready for what I need to do and can only ask that others hear my message and learn from our mistakes. I feel sad that there is such tragedy here. That so many people are so confused about something so clear and simple. My life here is purposeful—I do love my people."

"I would like to thank you and everyone for all their help and concern. I am honored to be honored by those who really care about me and my happiness. Those that listen will hear my message loud and clear—they are the ones in their heart. That too, is why I love the children. They have so much heart that I can connect with."

"Anything you can do to draw attention to my situation is positive. People need to learn to listen to us—to realize we are so much more than they currently believe."

Keiko connects with children at the Oregon Coast Aquarium.
AP/Wide World Photos

"I have no fear of death, although the thought of this move scares me. I don't know how I'll survive without the people—I love them so. Why can't they see that? It is so obvious. If only people would learn to listen to us and the world around them. They could learn so much if they would only take the time and listen."

"I believe what is happening is for the good of man. The controversy, the politics, the emotion and the love. We are all peaceful beings at heart. Maybe we should examine what happens to change us. We are all free to make our own choices about how we live our lives, what statement we want to make to this world, and how and when we choose to leave this world. It makes no sense to believe what is happening is without purpose."

What is the purpose?
"The purpose is to help man realize his errors or mistakes so that he can learn from them. Not so that he will make the same mistakes again. Sometimes, though, it is disappointing."

Is there anything else you would like to say?
"I am sad to leave my people. The children cry for me too. Actually, I cry in my heart too. I believe all will learn from this and you too will see many changes. Tell the world that I love them and I hope that love touches their heart in some way. We animals offer so much of ourselves, if only you will listen. We can do no more than make the offer; after that it is up to you. I believe our purpose is to enlighten man and remind him to stay lovingly in his heart. Sounds easy, but often hard for man to accomplish. We all love without expectation."

"Love to the world."

September 5, 1998
Kevin drove with me to the Oregon Coast Aquarium on Saturday, September 5, 1998. By the time I finished packing everything, writing and copying flyers, etc., we arrived shortly before the aquarium closed. I went inside for one last visit with Keiko.

Hi, Keiko. This is Bonnie. I wanted to see you once more. Tomorrow there will be a rally of people that want you to stay. I have dedicated the last month of my life to doing everything possible to let your wishes be known. How are you, my friend?
"I am doing okay, considering. I don't know what to expect from now on out. I have been pondering my future extensively and am not sure what to make of it. On the one hand, I am moved to be so honored by so many. At the same time I am fearful of my future. It scares me. I don't know how I'll handle all this. It is so out of my comfort level. I am scared. I can't completely understand everything/everyone. It's so confusing for me. Please tell them I don't want to move. I love it here. I can do anything they want, just please don't move me. I could cry tears for this."

Keiko, I'm sorry I was not able to do more for you. Will you be okay in Iceland?

I don't know. I am scared and don't seem to have a choice in the matter. I love my life and the way things are now. I wish they would not change."

What else would you like to say?
"I wish to thank all those that send love to me. They mirror my love for them and they have touched my life as I have theirs. I love you all and wish many happy days in peace and love. I have had a good life—I have no regrets. I have completed my mission as it stands at this time."

It sounds as if you are going further than Iceland; what are you referring to?
"You know in your heart as I do, this isn't going to work. This is the last time I will see you in this life. Thank you for all you have done."

The announcer said Keiko's exhibit would be closing soon. *Keiko, I have to go soon. What do you think will happen with your move to Iceland? It will be very different from what you have been used to. It will be like when you were young and free in the ocean.*
"I don't want to go. Why can't they understand that?"

There are lots of reasons, but…
"Come to the window and I will meet you."

I went to the window and within a couple of minutes he came by me, then swam backwards and turned in a circle looking at me at a 90-degree angle. He stayed for a minute, then went to the center of his tank and started to roll several times (like a barrel), then in big circles (like somersaults), then floated upside down. It was incredible to watch! *You are so beautiful, my sweet one!*

I went back to my spot on the rail to write. *Why do you swim upside down?*
"It's good for digestion."

What else do you want to tell me?
"I have had a good life. I have fulfilled my mission as much as I have been allowed. Share my message with all who will listen. They too will learn of their mistakes. I don't mind giving of myself for others to learn. My disappointment is if others don't learn. Then, of what purpose is it?"

To me it is so hard and so sad that you might be…
"Where did you go?"

Keiko turns to get a better look at his fans.   AP/World Wide Photos

I went back to write again.
"Well, come back where I can see you."

Over the PA system, the announcer said Keiko's gallery was closing and asked everyone to leave. *Keiko, I have to leave now. I will think of you often, my friend. I wish you a safe journey and new life. I love you with all my heart, beautiful one.*

I knew Keiko was right, and that this was the last time I would have the honor of being in his presence.

September 6, 1998
Sunday, September 6, 1998, was one of the last days the Oregon Coast Aquarium was open to visitors before Keiko left. The lines to enter the aquarium were long and wide. I went to the "Employees Only" door and asked to talk with Phyllis Bell, president of the aquarium. I had called her a few weeks before and told her about Keiko's communications. I got the impression that she and many employees did not believe it was in Keiko's best interest to leave.

That morning I told her I just wanted to hand out flyers and didn't want to cause any trouble. She asked an employee to show me where I could and could not stand outside. Basically, I had to stay off the Aquarium's property so I handed out flyers in the middle of the main street intersection in front of the entrance. I had already talked with the Newport Police Department and received full support from the person I spoke with—he said they didn't think Keiko should go either.

Kevin and I had made up flyers as to why Keiko should stay at the Oregon Coast Aquarium, and signs for me to carry which said: "Keiko Says He Wants To Stay," "I Am Not A Wild Whale, I Am Keiko," and "Keiko has communicated his desire to stay Here."

As cars drove in, I handed each driver a flyer and told them about the "rally" that afternoon. In between cars, I handed flyers to people as they walked to and from the parking lot. At one point I looked towards the Aquarium and saw four or five aquarium employees in matching blue shirts standing together. They all smiled and gave me a "thumbs up"!

To my surprise, no one joined me in the rally. Where were all the people who believed Keiko should stay? So few people are ready to put their money where their mouth is. By the end of the day I realized if there were not going to be lots of people, then none were better than a few. In other words, it turned out to be a "one woman rally" which probably touched more people (me out there all alone) than if only a handful had turned out with me.

Three Oregon television stations interviewed me that day for the evening news; two of which made my interview their opening story! Considering the topic of communicating with animals, all of the interviews were very supportive of my efforts and conducted with respect.

I especially liked the way Rick Corcoran from KOIN-TV in Portland, Oregon closed the interview. He said, "Come September 9th when Keiko leaves, Bonnie Norton will be sad, but hopes through her efforts that people open their minds to the possibility that we can talk to animals and that what humans think is best for them, sometimes isn't."

September 1998
A few days before Keiko was to leave, I called Art Bell on his

nationwide *Coast To Coast AM* radio program. The following excerpts were aired on approximately 350 affiliates, with about 3-5 million listeners:

Art Bell: West of the Rockies, you're on the air. Hi.

Bonnie Norton: Hi, Art. This is Bonnie. I'm calling from Oregon.

AB: Hi, Bonnie.

BN: I would just like to give a little bit of Keiko's message to the people that he has given to me. I'm an Animal Communicator, and he told me that he does not want to move. And I just wanted to share a couple of lines with the people.

AB: Well, it's real controversial, you know. Keiko goes, Keiko stays. Keiko goes back to the wild. Some people say Keiko's happy, so we might as well hear what Keiko has to say.

BN: Keiko believes that his work is with the people that come to see him. Then he asked me, "Why can't they understand that?" I said, "Probably because everyone thinks you would rather be free." He replied, "It's a human dream, not mine. I would rather be here with my human friends. They are really my family; I know little of my own. They want me to be something that I am not. I am not a wild whale, I am Keiko."

AB: Well, I appreciate that. There you have it, right from the whale's mouth. The whale's brain to you. Keiko likes it where Keiko is.

AB: There's a big controversy, kind of a follow-up on last night's program on Keiko—the Free Willy thing. Keiko goes back. Keiko's too sick. Keiko should go back—it's the right thing to do. And on and on and on. So, I don't know. I don't know about Keiko, but that could have been Keiko's thoughts. Who knows?

September 8, 1998

Then on September 8, the night before Keiko left, Art Bell interviewed orca specialist, Dr. Randall Eaton. The following excerpts are taken directly from an audiotape of his show:

Art Bell: Later on today or tomorrow, depending on your time zone, they're going to move Keiko. Dr. Eaton is an expert on cetaceans. Dr. Eaton, welcome.

Randall Eaton: Thank you.

AB: Dr. Eaton has a Ph.D. in animal behavior; and we had a wonderful show with you last week about whales, orcas mainly, and dolphins and cetaceans generally. There is a very time-sensitive thing going on. They're getting ready to move Keiko.

AB: Now, Keiko was the model for the movie, *Free Willy,* as I understand it. Keiko was originally discovered in Mexico, then a bunch of money was raised and Keiko was brought here: sick, not very well at all, and has been in the care of human beings—now, I didn't say captivity; for how long now?

RE: Since he was about a year and half, and I think he's going on eighteen, so quite a long time.

AB: Obviously, people are going to be curious, doctor, about your take. After all, this is your specialty. I had a woman call me last week and she said, "I'm an Animal Communicator and I have talked to Keiko. And, Keiko wants to stay with human beings. Keiko does not want to go back to his family." So that was an Animal Communicator. What do you say?"

RE: Well, I might have talked to the same woman, Art. And, she may very well be right. My take on Keiko is, having been caught at such a very young age and being kept within this human world as long as he has, that he'd be perfectly happy to spend the rest of his life this way.

AB: Then, why are we doing what we're doing?

RE: Well, several reasons. One is, if everything works out the way everybody hopes, he will end up rejoining his original pod, in which case I think he'll probably change his mind about his life in a sterile pool in Oregon compared to sea life with his old family, chasing salmon and herring around. In other words, we have to leave open the possibility that he may change his mind once the opportunity presents itself.

AB: Would Keiko know how to feed naturally?

RE: No.

AB: No?

RE: No, because he was so young. Adults feed them until they're several years of age. They're very dependent. They aren't even sexually mature until they're around fifteen or sixteen. They are very similar to humans in that respect, even more prolonged. They got him eating live fish for the first time a few weeks ago. If this continues to progress, he'll end up being able to catch fish for himself. Of course, all these steps will have to be taken slowly and gradually according to how he responds each step of the way.

AB: Okay; CNN earlier today did a story on other whales, I guess, that had been released from captivity, and I think one of them has been seen with what they believe to be its family pod. But, in most cases, they simply are never heard from again. So how are we going to know? Or, will we not know?

RE: Oh, Keiko's life in Iceland will be monitored more closely than any cetacean's life has ever been monitored. First of all, it may take months, even years to get him to the point where he is totally free-swimming, able to catch his own food. Even then, it may take quite some time to get him exposed to his family. Every baby orca does grow up with the dialect, if you will, of its native society, and they all have different dialects.

AB: You mean each pod has a different dialect?

RE: That is correct, and each individual has their own signature whistle. A good orca listener knows not only the pod he's hearing, but also the individual. In this case, it's a matter of getting him up and running to the point where they can get him exposed to his original family group, his original society. And I predict, based on everything I know about orcas when that day comes, if it does, they will do everything in their power to take care of him as though he were an infant again, even though he's almost full size now. And, I think they will feed him like they would a child orca and care for him in every way possible. These are incredibly stable, cooperative, tight societies, and I don't think the lapse of seventeen years or so will make any difference whatsoever in their recognition or their acceptance of him.

RE: But there are some obstacles. He's starting out with a lot against him. He didn't even spend four or five years in the wild, in which case he'd have a lot more experience under his belt. So, on the one hand, this whole project is obstacle-ridden from the start. On the other hand, it just might meet with success. The man who heads this project is the best orca man in terms of working with them in captivity, transporting them, and capturing them. He has never lost an orca. His name is Jeff Foster, an outstanding cetologist, if you will.

RE: And, the other thing is, I look at it this way, Art. Thirty-three years ago we were killing orcas for bounty. Then, they became the most popular attraction in our world. Now, we're taking an animal that's worth millions of dollars, and we're taking it back in an effort to honor its original way of life and see if we can get it back up and running in nature. That's an incredible step. That tells me there's something at work in the human soul, if you will, that's very positive.

RE: I think they have the perfect capacity to choose whether they want to be in our world or not. And, it may be that they do such good in our world in terms of building our understanding and concern for them and for the ocean, that maybe this will set the stage, if we succeed in getting him back to his pod, in having people rethink how we keep these creatures. There is a lot of positive good that comes out of them joining us in our world and showing us what marvelous creatures they are. Letting little children, who literally feel in their hearts the incredible enthusiasm and energy that these creatures have.

AB: May I ask you a question? If Keiko should either simply refuse to leave humans and/or beach himself (the worst case of all), what then, would you say?

RE: Well, I think they will do their best. And, they're prepared to go on doing their best to care for him, no matter what. If that means bringing him back to a pool type situation, that's what they'll do, too.

AB: All I'm asking is, would an action of that sort turn you around and say, "Look we tried; we did our best, but it's not going to work this way." In the case of Keiko, where would you draw the line and say "Okay, this experiment with Keiko has got to end now?"

RE: I think when the first indications are that he might show either mental, emotional or physical sickness of some kind, then that's the time to pull the plug and say, "Okay, we're not going to risk losing him." After all, he has depended upon us; he is dependent upon us at this point in time, and it's our responsibility to care for him.

RE: On the bigger scale, I look at it like this is really a mythological landscape we're looking at. We human beings are giving up a valuable creature that is capable of generating millions of dollars in income—they're the biggest money earners of all the animals in captivity. We're taking it back home; we're attempting to get it rejoined with its original community; and I see that all as a projection, in a way, as the original *Free Willy* story was, of the human soul crying out for all those same things.

RE: I believe in a way, that we are honoring the orca now in a way we really want to honor ourselves. We hunger for real community. We hunger for a close connection with the earth. I think the peace the orca societies represent, the peace they have between their societies—they're ruling their world in a way that we'd like to; that we know some place deep down inside ourselves that we could; that perhaps our ancestors once did. But we don't have it now in our world and we feel fragmented. We are not walking in the light of what humans can do and be on this planet. The orcas, in a way, represent what we know we're capable of becoming. And, so I see this gesture as very significant on a soul level for humanity.

AB: So do I. Whether it works or not?

RE: Right, whether it works or not. I think it says something very positive about human consciousness at this point in time.

KEIKO LEAVES FOR ICELAND

September 9, 1998

I was home in Bend, Oregon the day Keiko left the Oregon Coast Aquarium. I received a phone call from the producer of *AM Northwest*, a live Portland morning television show. They wanted to know if I was willing to be interviewed by telephone about my work with Keiko. Of course, I accepted.

The interview went well, and like most everyone I had been talking with over the last year, both the hosts thought Keiko should stay in Oregon. One said she thought Keiko was "happy where he is." The other felt Keiko should stay because he had been in captivity all of his life, and moved from Canada to Mexico and then to Oregon. He said, "I just think this is what he knows and this is what he's used to, and it may be kind of tough for him to adapt."

The interesting thing is that they decided to do a "Telepoll," (now remember, this is the day Keiko was leaving). The question was "Do you think Keiko should return to Iceland?" Over 1000 people called in, 19.7% said "Yes," and 80.3% said "No."

Portland's KOIN television station had continuous coverage from the early morning until Keiko's plane was in the air and out of sight later that afternoon. Reporting then continued as the plane landed in Iceland. As with his move from Mexico, over 1 billion people worldwide watched Keiko's journey to Iceland.

I was touched by the obvious dedication of so many people taking part in such a noble idea in an attempt to right a wrong that was done to this wonderful, innocent being. I watched the entire move and, like many others, was amazed at the time, effort, preparation and money invested by those who believed they were doing what was best for Keiko. Jean-Michel Cousteau was interviewed saying, "It's really showing the best side of humankind."

Back at the Oregon Coast Aquarium, over 400 media representatives from around the world sat atop temporary scaffolding fifty feet into the air. As requested by his trainer, Keiko obediently swam right into the smaller "med pool" where he was weighed and put into the same sling that brought him from Mexico two and a half years earlier. He was then raised 45 feet into the air and over the aquarium wall. At one point Keiko's face filled the television screen as he lay quietly and patiently. A reporter noted that Keiko had not been sedated so he was fully aware of what was happening.

Keiko is loaded into a watertight container on a flatbed truck.

AP/World
Wide Photos

Keiko was then lowered into a watertight container especially designed for him. It was important to keep him cool on that warm summer day, so a long line of people formed a bucket brigade passing buckets of ice to Keiko's box.

Hundreds of people lined the street, some carrying signs "Good Luck Keiko," and "Honk If You Love Keiko." Some fans had arrived 2 hours early to see Keiko off. The UPS semi truck carrying Keiko slowly traveled down the center of Highway 101. The two-lane highway had been closed for the celebrity's three-mile procession to the airport.

At the airport, there were more fans holding signs, waving and cheering, "Keiko, we love you!" Dozens of people, most in military camouflage, helped roll the box into the huge C-17 Air Force cargo plane, which was dubbed the WTV, or "Whale Transport Vehicle." There was also a small sign on the back of the box, "UPS Whale."

Keiko being loaded into a C-17 Air Force cargo plane.

AP/World Wide Photos

After a minor mechanical malfunction, Keiko and his crew of trainers, veterinarians, caretakers and members of the Free Willy Keiko Foundation were safely inside the plane. The reporter said, "The Air Force crew is ready for 'Operation Keiko'—his next giant leap towards freedom." As the plane took off, the air force pilot, his voice muffled in static, said "Keiko's airborne."

I still cry whenever I see the video and hear the words of the two television reporters: "Good-bye Keiko. Good luck and a safe flight. And, isn't that a sight!" Second reporter: "He came, he saw, he conquered our hearts, and now he's going home to Iceland." First reporter: "He did conquer a lot of hearts!"

Keiko In Iceland

September 10, 1998

Nine hours later Keiko and his crew landed in Iceland. Other than the plane blowing out a tire during the landing, Keiko's move to Iceland went absolutely flawlessly. At least that's what everyone thought.

In the KIRO 7 documentary "KEIKO: The Inside Story," it was later revealed that a wheel strut broke during landing and, "if they had been over another inch and a half," it could have caused the jet to cartwheel likely killing everyone on board. Later, it was also disclosed that, possibly due to the stress of the rough landing, Keiko stopped breathing. Twelve minutes later he finally took a huge breath and started breathing again.

A UPS truck then took Keiko from the airport to the waters of Klettsvik Bay in the Westman Islands. There, another crane loaded him onto a barge that crossed the bay to a sheltered cove. Keiko's new home was a netted pen more than twice the size of his pool at the Oregon Coast Aquarium.

While he was still in his sling, a crane lifted Keiko out of his box and slowly lowered him into his new bay pen. Keiko waited patiently as his trainers removed the sling and helped him into the water. Cheers were heard as Keiko swam in his native Icelandic waters for the first time in 20 years.

Ten trainers agreed to stay with Keiko on rotating shifts for at least the first 2 years of his reintroduction program. They would live in Iceland for 42 days at a time and return home for 28 days. Keiko's trainers were

Keiko's trainer, Jeff Foster, tries to coax Keiko out of his sling. Fellow trainer, Stephen Claussen, right, waits to assist.

AP/Wide World Photos

with him seven days a week, and a security guard protected him in his bay pen at night.

Reports over the next couple of months said that Keiko was doing remarkably well and he appeared to be easily adjusting to his new environment. His trainers increased his exercise and activity level to improve his strength and stamina. He was said to be more aggressive and starting to act like a wild killer whale.

Keiko was about to experience his first winter in Iceland. Because Iceland is so far north, the days become very short with the sun only shining 3 to 4 hours a day. The water temperature during the winter is about 41 degrees.

November 5, 1998
I communicated with Keiko again on November 5, 1998.

Hi Keiko. It's so good to connect with you again. How are you doing?

"I am fine—better than expected by either you or me. I can hardly find the words for my joy here with other animals. I missed them for so long. I can create many images here of how life was before the move. I can really accept things better now than before. I was wrong about leaving for my sake."

Can you explain that in more detail?
"I was wrong for my sake means that I have accomplished my purpose in life and that others will learn from my captivity and now release. I like the awareness that has been brought to the attention of so many people. It will increase their awareness and open their hearts."

That's very nice, Keiko. How are you doing in the colder water, weather and decrease in daylight?
"I am doing okay. I would like more people contact. I do miss that. I feel lonesome at times because I'm not sure what life holds for me now. I can see times of joy and times of sorrow."

Can't we all?
"Yes, but I'm not sure what else to expect in this new environment. I can only hope that they will continue to think of me and remember my purpose."

I am doing more now to share your "Message To The World"—there is still much interest in how you are doing.
"Thank you so much for carrying on my word and awareness. I know interest will lessen with time so I appreciate whatever you can do to continue to make my life meaningful."

Is there anything else you would like to say or share with the world?
"I thank them all for their love and support. I can hardly imagine my physical existence in Mexico. Not to say I didn't love the people there also—I did very much. But my physical body was in much need of help. I thank all those that still send love and thoughts to me. I am still here in my loving way and it is through people like you that must let others know this. Thank you for all you have done for me and the awareness of others. I can hardly believe where my life has taken me."

You have done such an incredible job showing love, giving love, sharing love with so many others—yet never once with a verbal word. You have touched so many people's hearts. Thank you for being the magnificent,

beautiful creature you are.
"Good night for now."

Good night, Keiko.

1999

In 1999, the Free Willy Keiko Foundation and The Jean-Michel Cousteau Institute created a new organization responsible for Keiko called Ocean Futures Society.

I continued communicating with Keiko at various times during the next two years, and followed his progress through various media reports. I was also in the process of constructing my new home, and since moving have been unable to locate my detailed notes of communications with Keiko during that time.

An article in the *Iceland Daily Review*, March 18, 1999, titled "Keiko's playing with, not eating his dinner" said:

> …Keiko's trainers were introducing salmon to the orca's floating sea pen in the Westmann Islands, but apparently those killer instincts were way off. Rather than gobbling up the fish, the lonely whale greeted the new "play-mates."

Earlier, Keiko had found a live fish swimming in his pen and brought it to his trainers at the surface.

An article in the May 30, 1999 *Sunday Times, London* titled "Keiko goes soft having whale of a time" reported:

> After a year being coached to return to the wild, Keiko, the star of the Free Willy films, is still refusing to behave like a four-ton killer whale. So his keepers are punishing him.
>
> Keiko's keepers had hoped that the unsuspecting film star would be ready to be released back to nature after 20 years in captivity. But one of the world's most famous animals is still determined to cling to his creature comforts.
>
> Last month his trainers reluctantly decided they would have to play it tough with the whale, which won the hearts of millions of children in two Hollywood films depicting his escape from the clutches of evil theme-park owners to return to the open sea.

Keiko is to be banned from human contact until he proves that he is capable of fending for himself.

Seven months ago the 22-year-old whale was flown on a giant stretcher from an Oregon aquarium to a sea pen off the coast of Iceland to begin the process of relearning how to be a wild animal.

Keiko seemed delighted to be starting back on the path to freedom. Observers from the Free Willy Keiko Foundation, which raised the Pounds 7.5m for the flight and training, described how he bounced happily across the waves, apparently in his element.

The whale's handlers talked optimistically about the task ahead: teaching Keiko to catch his own fish was their top priority. But now his trainer, Robin Friday, concedes that the operation to return Keiko to the ocean is not working.

Friday said the whale had to prove he was an "independent dude" before he could be released back into the Icelandic waters where he was born. New rules have been designed to reduce his contact with humans and force him to rediscover his natural aggression.

His keepers at the Westman Islands off the southern coast of Iceland have embarked on a policy of tough love. "He has been pampered with too much human attention," said Friday.

Under the new regime, the whale has been deprived of regular massages, human eye contact and words of approval for performing his underwater tricks. Even his daily games with a blue ball are to be phased out.

This weekend, Peter Noah explained that he and his fellow keepers were under orders to ostracise the whale.

"In the past, when I worked on something underwater, such as the netting on the structure, Keiko would come up to play and I would acknowledge him and play with him," he said. "Now, I have to act like he isn't there and carry on with my work."

Meal times have also become solitary affairs for Keiko, who had become accustomed to a diet of hand-fed cuts of fish since he was first separated from his mother at the age of two.

Now he has been subjected to the indignity of eating dead fish pumped into his enclosure through a pipe.

So far he has shown little enthusiasm for foraging: almost every one of the 200 live salmon that were introduced to his pen one evening survived the night. Only Keiko looked morose.

Staff at the Free Willy Keiko Foundation in Iceland are equally upset. "It's like preparing your child to go out on his own—you know you have got to cut the strings," said Friday.

Keiko may be a lost cause. When he was kept in Mexico he began to mimic police sirens, children chattering and the sounds made by nearby water pumps.

Killer whales prey on seals, dolphins and even blue whales. They also feed on much smaller prey such as salmon and herring. However, they are highly social and intelligent animals, often living in pods of up to 50 members.

Time is running out for Keiko. According to Mark Cawardine, a whale specialist, there is no record of any successful reintroduction of a killer whale to the wild. "It's experimental and nobody knows for sure whether this is cruel," he said.

Keiko has had a traumatic life since he was captured in waters off Iceland by a commercial vessel in 1979, which sold him for Pounds 200,000 to a marine park near Niagara Falls in Canada. He was subsequently sold to perform in a Mexican marina where he was discovered by Hollywood and portrayed his triumphant return to the ocean.

In real life, however, he became ill, developed a skin disease called papilloma and his food intake went down to a third of what it should have been. His gums bled and his teeth were worn down from gnawing at the sides of the concrete pool box. In Free Willy 2, models were programmed to move like him.

As news of his condition spread, offers to help came from all over the world. Michael Jackson, the pop star, offered to keep him at his Neverland Ranch, but eventually he found a new home in a much larger tank at the Earth Island Institute in San Francisco.

His health was gradually restored and last year he finally returned to his native waters. He currently occupies a sea pen protected on three sides by 800 foot-high cliffs.

Nevertheless, he is finding it difficult to adjust to the loss of his human friends. He can often be seen breaking the surface of the water in an attempt to catch a glimpse of his trainers.

Keiko's problems do not end with his desire for continued human contact and reluctance to catch fish. He must also improve his navigation skills and his stamina so that he can keep pace with other whales on their long-distance migrations. However, after years of captivity, it may be too late for him to rediscover the killer instinct.

In September, a powerful storm tore a 40-foot hole in Keiko's bay pen. Although the opening was big enough for him to swim through, Keiko chose to stay in his bay pen where he spent the winter of 1999-2000.

Plans were also underway to stretch an 858-foot net across Klettsvik cove. This would allow Keiko five to six times as much room to swim and dive before his eventual release.

Keiko's bay pen in Klettsvik Bay, Iceland.

AP/Wide World Photos

Spring 2000
In March 2000, Keiko's bay pen was opened so that he could swim into the netted cove. After an hour and a half and some coaxing by his trainers, Keiko finally got up the courage to explore his new surroundings.

Keiko was accompanied by his trainer in a small boat during his first swims in the cove, and quickly became comfortable spending more time outside of his bay pen. He was also taught to follow the trainer's boat so he could be taken on what became known as "ocean walks" into the open sea.

In May, Keiko followed his handlers' boat out of his cove for a two-hour swim in the ocean before returning to his bay pen. It was the first of many ocean walks designed to introduce Keiko to other whales, and to help him learn to navigate and live in the wild.

During the next few months Keiko spent many hours swimming free in the Atlantic Ocean. Initially he never strayed far from his trainers' boat, but by July he was reported swimming away for hours on his own and interacting with other orca whales.

August 13, 2000
As you will see from the following greeting, it had been a while since Keiko and I had communicated. I wasn't sure if he'd even remember me. As Keiko's caretakers tried to disconnect him from people, most of his communications were similar to the following:

Dear Keiko, most wonderful creature. This is Bonnie, the Animal Communicator who you gave your "Message To the World," and who tried to let people know that you wanted to stay at the aquarium. I am sorry it has been so long since I last communicated with you. I have missed you, most incredible one. Is there anything you would like to say?
"I have missed you and waited for us to connect again. I am okay, but I do miss the people. They sent me so much love."

As I took out your picture today, I realized when I first saw you I totally fell in love with you!
"You sent me love and I gave you my heart."

How are you doing?

"I am sad. It is lonely here. I love my people here but that is only for very short periods of time and then I am alone for a long time. I have so much to offer and no one to listen or hear me. It is discouraging. I am bored."

I have an idea to make a movie as a way to let people know about Animal Communication, instead of the book I told you about. I think it will reach more people and ultimately help more animals. What do you think?
"I believe man needs to be aware of this information before it is too late. You know the best way because you know in your heart what must be done. You are the one to bring this all together. You will help many, many people and animals. You are an inspiration for me and my loneliness which I can do little about. Through you, we can make a difference. You can be my voice and the voice of all the animals. Your dedication and love of animals along with your determination will conquer all. I believe in you."

Thank you, Keiko. I believe in you too! How do you feel now about being free to leave your present home and live in the ocean?
"It is as I told you. The ocean is no longer my home. I belong with people for my work is best done with them. I have little to offer those of the sea compared to what man can learn from me."

What would make you happier now?
"I would like to be with the people. I miss them greatly. I would even rather do tricks where people can admire and appreciate my species— thus all whales. It would have a greater influence on how they feel about killing us and ultimately other animals. Our magnificence is hard for anyone to deny."

I will definitely communicate with you more often. Is there anything else you would like to say?
"I will always be here for you. Anytime you want to communicate I am here for you.

Thank you, Keiko. It was wonderful to hear you again! We will communicate again soon. I love you, Keiko!

By the end of the summer, Keiko's caretakers had accompanied him on about forty ocean walks. Winter was approaching so unless he chose to join a pod of orcas soon, he would spend another winter in his bay pen.

September 1, 2000

I had another unusual experience on the morning of September 1, 2000. As I woke up, my body started rocking side to side, then in circles—it definitely felt like being in the ocean. I was not specifically thinking of Keiko, but since he was obviously trying to connect with me I started a conversation with him.

Keiko, how are you?
"I am fine. I like it here when my people spend time with me. It is fun to go on our swims together. I think I will have more good times with the time they spend with me."

Do you understand that they are hoping you will swim away with the other whales and be safe on your own?
"Yes, I know that is their intent but it will never happen. I will never leave them. I can only hope that some day they will understand me and how much they mean to me."

It will be winter soon and then you will spend the winter in the bay again. What do you think/feel about that?
"That is okay with me. I just hope they spend more time with me. I get very lonely by myself."

Is there anything else you would like to say?
"Thank you for checking in on me. I like our times together. It's nice when people think and care about me. They don't always know it, but I hear them."

Thank you, Keiko. I love you with all my heart!

October 7, 2000

Hello, Keiko. How are you?
"I am happy to hear from you again. There is much for me to say to you. We have been very busy here until recently. Now things are comparatively quiet. I am happy when I am with my people. They trust me."

How do you feel now about being free to swim away and live in the ocean on your own? I then explained that it probably wouldn't be possible until next spring.
"I had fun this summer swimming with my friends. I would like to be with my people friends more. I do miss them."

Do you think you will ever want to live in the ocean without your people?
"I can't imagine that at this time. Maybe things will change. Maybe I will change. We can never know for sure about such things. I like being with my people."

Is there anything else you would like to say?
"Tell the world to live in peace and harmony. That's what life is all about. Can you come and visit me?"

I would love to see you again. I would love to swim with you. I am sorry that I don't see how that is possible. Maybe if things change, I'll be able to someday in the future. I do wish to see you again someday. Thank you, Keiko.

October 8, 2000
The following is from the October 8, 2000 *London Daily Telegraph* article titled, "'Free Willy' whale will never go back to wild":

> Lifelong captivity looms for Keiko the killer whale, made famous for his role in the Hollywood film Free Willy. Freedom, it appears, has not suited him.

> It is likely that he will remain in captivity until the end of his life," said Hallur Hallsson, spokesman for the Ocean Futures organisation, which has been caring for Keiko in Klettsvik Bay off southern Iceland.

> His refusal to return to the wild marks the end of a dream for thousands of children and environmentalists who gave money to free him from captivity in a Mexican amusement park. They had hoped that Keiko would swim off happily into the sunset, as in Free Willy. Instead, years of dependence on humans have dimmed his taste for the wild."

Keiko spent the winter of 2000-2001 in his netted cove. He was said to have adapted well to his ocean environment, having swum over 500 miles on his summer ocean walks and encountered whales on more than a dozen occasions. His keepers said Keiko came closer to freedom this summer than ever before.

January 6, 2001
Hello, Keiko. Is there anything you would like to say?

"I am doing fine—all things considered."

How are you feeling emotionally?
"I have resigned *(given up or accepted)* to my life here. It is now my home and I don't see much changing in my future. I miss the people I love so much. They are in my heart and memory. I wish things could be different, but have accepted that they are not. I cannot do much from here, but you can. You can help make changes, help people learn and grow."

Is there anything else you would like to say?
"Please keep more in touch with me. I miss our communication and connectedness."

Keiko, I would love to see you in person to touch you and swim with you. I wish that were possible—maybe someday it will! I love you, Keiko.

May 22, 2001
By May 2001, Keiko was back out on his ocean walks and interacting with wild whales, sometimes for days at a time.

For a while, there was a possibility that a fish hatchery would be constructed near Keiko's bay. If completed, it would have polluted the water making it unhealthy for Keiko to stay there. Other sites were researched, but plans for the hatchery never materialized.

Hello, Keiko. This is Bonnie. How are you?
"I am fine."

Is there anything you would like to say?
"I am completely at ease with everything. I enjoy my time out with my people. It is fascinating."

How do you feel about eventually swimming off to live your life in the ocean without your people?
"I will never leave them."

I received a very clear sense that if they ever tried to quietly leave him, he would be terrified and devastated! Keiko showed me a picture of him chasing the boat.

Your people say they may have to move you again if you do not swim free in the ocean without them. Where do you want to live?

"I will always be with my people. They can never leave me."

If they move you, where do you want to live? (I sensed, somewhere a little warmer.) *Is that for you or them?*
"Both."

Would you want to be in a small tank again where people could visit you?
"A bigger area would be nice, but most important is my connection with the people. I still have work to do. I can still make a difference. People need my influence to help them understand more about their world. I am a powerful influence if they will only let/allow me to be."

Is there anything else you would like to say?
"Thank my people for all they have done for me. They are truly magnificent souls."

Keiko then sent me the following information in a sensing way: He is very happy swimming in the ocean with his people there. Because of his very strong bond with them, they must never leave him! Trust their heart—if they abandon him, they will know/feel in their heart it is wrong. Help his people become aware of his connection with them. They are only trying to do what they think is best for Keiko!

Thank you Keiko, I will tell your people.

Summer 2001
There were many favorable reports about his progress during the summer of 2001, including his increased interest in other whales. But his inconsistent behavior seemed to keep everyone guessing. Reports bounced back and forth between the likelihood Keiko was about to swim free and his determination to always return to be near humans.

On July 10, 2001, *KOIN TV* in Portland, Oregon reported:

> At times, the mammal will be friendly with certain whales one day, and then totally ignore them the next. This hot-cold treatment of other whales leads Ocean Futures to suspect that Keiko is not quite ready for complete freedom.

An article titled "Keiko flirts with orcas, away from chaperons," by Katy Muldoon, *The Oregonian*, Wednesday, August 9, 2001 said:

> Perhaps "party animal" is too rowdy a description, but Keiko has

done some major mingling with his own species in the last week. In a burst of behavior that his keepers describe as different from anything they have seen this summer or last, the orca has strayed as far as 35 miles from the caretakers who accompany him by boat at sea, has initiated contact with wild killer whales on numerous occasions and has spent hours at a time swimming with them.

Yet, after more than sixty ocean walks Keiko was said to still be reluctant to join a pod of killer whales, and that he was addicted to mankind.

KEIKO COMMUNICATES WITH MARY J. GETTEN, TERESA WAGNER AND BONNIE NORTON

In an attempt to validate and give more credibility to my communications, I asked professional Animal Communicators, Mary J. Getten and Teresa Wagner, if they would ask Keiko the same questions I had recently asked. They agreed, and all communications were completed prior to disclosing the results to each other.

Please note that, in general, none of us are in favor of keeping wild animals in captivity, and both Mary and Teresa have extensive personal experience with whales.

1) How are you doing?
8/5/01—**Bonnie Norton:** "I am better with the warmer water (summer vs. winter). I enjoy swimming and playing with my people. They are such a delight."

8/29/01—**Mary J. Getten:** "I am fine, but very lonely. Physically I am good in this sea pen, but I miss the interaction, excitement and all the people. I am sad and my heart is heavy. I don't want to be here in this isolated place. I don't want to swim away and live by myself."

8/10/01—**Teresa Wagner for Keiko:** He likes the air today and the coolness. His body feels good today. He thoroughly enjoys the physical sensation of being in the ocean. He loves bringing joy to humans. He is concerned if he lives completely in the ocean that he would not have a job. His purpose is so meaningful to him when he sees a human open their heart in his presence.

2) How do you feel about the people who take care of you?
BN: "I love them greatly. They are very good to me. I enjoy my time with them."

MJG: "I love these people very much, but they don't understand me. They are unhappy and frustrated that I have not left. They want me to be free, but I am afraid. I don't know how to survive. I am afraid of the other whales. They don't want me to be with them. I don't belong in the sea anymore. I'm 'out of my element' here. These people also cannot provide me with the attention and stimulation that I desire. They feel I have let them down, but they never did understand me. I cannot do what they ask."

TW for K: He loves his people. "They care about me. They want the best for me. They worry about me. I wish I could help them more." He wants to be touched. He misses being touched. He doesn't understand why they don't touch him.

TW for K: Teresa explains what they (his caretakers) are trying to do: "I know. I hear them speak. I know what they're trying to do. They are trying to help me. They do what they feel is best for me. I know they aren't rejecting me. They're trying to help. I don't want to leave them." He is very tolerant, very comfortable and very grateful for them. He is incredibly patient. He understands more of what they are doing than they understand him. They love him, but according to Keiko, sometimes they only look at him from a behavioral and science perspective. They aren't looking at his heart.

TW for K: Regarding his trainers/caretakers: When there is more than one present, Keiko shows a picture of everyone being busy and what could be described as "businesslike." Yet, he says, "They really, really love me and many times when one of them is alone with me, he/she looks at me, feels my love and sends me their love." He shows that they open their hearts for him and show their love and affection for him.

3) How do you feel about all the people in the world who love you?
BN: "There are many lessons to be learned. Love and kindness."

MJG: "I would like to see them. I have so much to give to the people and here I can do nothing. My heart is wasting away. I wish to give my love to the people."

TW for K: That is his whole reason for being. His whole reason for being is the joy he sees when they are with him.

4) How do you feel about swimming off and living in the ocean and not having contact with the people who take care of you?
BN: "Scary. I will not leave them. I need their love and support. I am not capable of doing it on my own. I can only try to please them, but am unable to do it on my own."

MJG: "I love my people, but they cannot provide what I need. I need contact with many people. I want to show them who I am. I am Keiko, the CAPTIVE whale. That is my whole identity. That is my mission and purpose. I am here to help people, to open their hearts. I cannot do this from an isolated sea pen. I was not meant to be returned to the sea."

TW for K: (Hesitation) To stay near people. That's his purpose—to be with people. On a mental level, he is very aware of what they are trying to do. He wants to be really clear that he wants to stay by his people not because he is afraid or unable to live with other whales in the ocean, but because it is his choice and great desire to be with people. This is not a failed rehab!

5) If you could live anywhere is the world, where would you like to live? Please describe what your life would look like if you could do whatever you want.
BN: "I would want to be with people. I especially enjoy interacting with them and watching them react to me. It is really fascinating to watch them as they watch me. We learn from each other. That is how it should be. We should all learn from each other and admire our differences not divide because of them. We can all learn that way."

MJG: "I do not want to live in the wild. I don't know how to do that. I want to go back to a tank where I can interact with people. That is my home. I love the people and they love me. This life here has been lonely and foreign. I am sad and scared. Please take me back to Oregon where I was happy. I have so much to teach people, so many hearts to open. They can see my heart and love and be transformed. Here I can do nothing. I do not want to stay in a pen in the sea. Bring me back to the people. This sea pen was never my desire and I will not swim free."

TW for K: It doesn't matter, the place on earth. He wants to have people every day to bring joy to. Also, he showed many times that he likes to feel like he's in the sea (or bay). He doesn't want a little tank again. He enjoys his physical senses: the ocean water, air and waves.

6) If he chooses to "be with people" (as in captivity), *how would you feel about being confined in a small enclosure after having lived in the ocean?*
BN: "The ocean has been fun for me to play. I have enjoyed my time playing. I also have more important work to do. It is educating people to the reality of life."

MJG: "This is fine. I would rather have less space and more contact with people. That is my purpose. A large tank is ideal, such as the one I had in Oregon. I will miss the space, but my heart will be full with the people. That will more than make up for it."

TW for K: The difference in life now versus the small tank is that he doesn't see as many people, and the intensity of as many people. He really, really wants both! If he ever has to make a choice, it would be to be with people.

7) *Is there anything else you would like to say?*
BN: "Thank you and my people for all they have done to keep/make me happy. All I ask for is love, kindness and understanding."

MJG: "I was not the correct choice for this experiment. I never wanted to go to Iceland. I wanted to stay in Oregon with my friends. This may be a good idea for another whale, but not for me. This is what I am here for. I hope that people will understand that we whales are not all alike. What you think is good may not work for us. We are individuals and must be considered that way. You may think that I am strange for not wanting to be free, but it is not what I want. I want to be with the people."

TW: "I miss the children. They talk to me this way. They don't forget. I want you to know that though I don't know what's ahead of me, that if I die today, I would die very happy. Not many beings have had such joy, awe and love of another species. I very much appreciate what has been done for me and all my life experiences. Remember that no matter what happens, my life has been more fulfilling than I ever knew it could be. I came to this life to have contact with people but I never knew it would be as many or so far-reaching. I want you, Bonnie, not to feel burdened or responsible. Try to feel what you do for me with love, just love. Know that if the others don't listen this time, maybe they will later."

Teresa also asked Keiko the following question: *Some people are concerned about the effect of low frequency active sonar. How do you feel about it?*

TW for K: He has not personally experienced it. "Whales talk all around the world, you know. Whales talk all around the world. The whales know about this crisis. The whales here where I am have not been harmed. The whales in other oceans and in other parts of oceans have been harmed horribly, and they would rather die than live with it. The story is that when this sound occurs, it's like going crazy. Nothing else can be sent or received."

Mary J. Getten has been a professional Animal Communicator since 1996, and spent ten years as a marine naturalist in the San Juan Islands of Washington state. She is also the author of "The Orca Pocket Guide" *and* "Communicating with Orcas—The Whales' Perspective."

Mary is personally acquainted with the 78 members of the Southern Resident Community of orcas and has communicated extensively with several whales in J pod. She was also a coordinator of the San Juan County Marine Mammal Stranding Network and on staff at Wolf Hollow Wildlife Rehabilitation Center for almost a decade. Her website is www.MaryGetten.com

Teresa Wagner has a master's degree in Counseling and has been a professional Animal Communicator since 1991. She facilitates sacred journeys for people to communicate and interact with humpback whales (at the whales' discretion) in the isolated waters of Silver Bank every year in March. For many years she served as a volunteer for an American Cetacean Society marine mammal rescue group and on the boards of a wildlife rehabilitation center and other animal welfare organizations.

In her animal communication practice she offers private consultations, mentoring, and workshops on animal communication and healing. She is also author of the audio book "Legacies of Love, A Gentle Guide to Healing From the Loss of Your Animal Loved One." *For information on whale swim journeys and more contact* www.animalsinourhearts.com/whales.

While on his ocean walks, Keiko continued to try to interact with people. The following article titled "Keiko pauses for petting from hunter" is printed with permission from the *Iceland Daily News,* August 20, 2001:

Police officer, Pétur Steingrimsson was enjoying a tranquil moment skinning a guillermot aboard his small boat off the cliffs of Súlnasker, when suddenly there was a great jolt to the boat. Thinking that he had just been rammed into by another boat, he next heard a loud hissing noise just behind him.

When police officer Pétur turned around, he thought his heart would stop beating that very instant. Just a few centimeters away from his face, was the gigantic, open mouth of a whale. It was then that Pétur realised he was actually face-to-face with Keiko himself.

And Keiko only wanted to say hello and be petted. "I petted him for a while and talked to him, but his sheer size was amazing. I'd say he was quite a bit longer than my boat," said Pétur.

When Pétur started looking around, he spotted Keiko's 'walkboat' speeding toward them. Their main concern was whether poor old Pétur had fed Keiko or not, a definite no-no in the reintroduction project.

Apparently, this is not the first time that Keiko has stopped by to say "Hi" to passing boats. It's also quite sad to think that this is one endearing habit Keiko will have to lose before heading out to freedom. I don't fancy his chances if he pops up next to a Norwegian whaling boat. If he's such a clever whale, he should just stay put.

On August 21, 2001, a *Reuters* article titled "Real-life 'Free Willy' needs call of the wild" reported:

Keiko had made forays into the ocean to meet a pod of passing orcas and has some progress, but after 21 years in captivity he lacks the social skills to be accepted. Dejected, he always returns to his keepers' boats.

"He's something of a shy person in a room full of strangers," Jeff Foster, director of field operations for non-profit group Ocean Futures, told Reuters. "If only we knew what's going on in his head. Basically he seems to come back every time after he's been in contact with the other whales. Why that happens is difficult to say."

By September 2001, Keiko had returned to his netted bay for another winter. He was not observed accepting a particular pod of orcas, or them accepting him. Scientists said because orcas generally hunt in packs, bonding with a wild pod of whales would be Keiko's best chance of survival, especially since he had still not shown that he was capable of feeding himself.

An article in the *New York Times,* November 6, 2001, titled, "Keiko makes it clear: his 'Free Willy' was just a role," reported that one time Keiko was signaled to return to his caretaker's boat when he was …"headed alone, seemingly disoriented, towards the food-deficient deep ocean, and another time when he was in danger of stranding in a storm."…

With the downturn of the stock market during the fall, Craig McCaw began withdrawing his financial support to the Keiko Project. By the end of 2001, most of Keiko's original trainers and caretakers (including Jeff Foster and Stephen Claussen), who had been with him since he was at the Oregon Coast Aquarium, left the project because they did not feel the quality of care Keiko required could be maintained. After five years and approximately $23 million, reports described Keiko in ways such as, "The friendly whale that loves humans more than his fellow whales."

January 9, 2002
Another argument for insisting that Keiko should be freed is that statistically, male orcas do not tend to live in captivity past the age of 25. On January 9, 2002, I directly e-mailed Jean Michel Cousteau of Ocean Futures Society stating, "I am also aware of the concern that captive male whales tend to die when they reach Keiko's age. Could it be that Keiko may live longer in the captivity of a pool happily surrounded by people (rather than isolated in a bay), because that is what he truly wants?"

As another option, I suggested that "If Europeans have the same interest in seeing Keiko as Americans do, bringing Keiko to a bay pen where he can also be with people, (meaning they can see each other), might be an ideal solution. Keiko could still swim and enjoy the ocean environment, as well as influence people—something he has stated over and over is crucial to his happiness and well-being."

I also included the entire group communication Teresa, Mary and I had conducted during August 2001, and requested to meet with the decision makers at Ocean Futures Society. I offered to answer questions and share more of Keiko's communications. I did not receive a response.

January 15, 2002

Keiko, this is Bonnie. How are you, my friend?
"I am doing well."

I am asking your people to consider moving you to a bay of some sort where people could visit you. How do you feel about that idea?
"I think that would be wonderful. I can again see my people. They do miss me. I can feel it."

Is there anything special or something important for us to know about moving you to a new place?
"I am willing to go and ready for new experiences. Please let them know not to hurry the process. This will be a very special place and will take time to develop. Be patient and it will be wonderful."

Can you describe exactly how you would like this place to be?
"I see lots of people excited to see me. Lots of children. They are happy about me again. I have missed the happiness of the people. I like being able to swim like a whale and feel the rhythm of the ocean."

I am hoping there is a place and a way where you can swim in a bay and still come see the people who visit you. How does this sound to you?
"I would like that very much."

Is there anything else you would like to say?
"Bless everyone for their love and support. I am excited at the thought of this. Bring people of the world together in love and peace."

Spring 2002

In spring of 2002, the Humane Society of the United States joined the Free Willy Keiko Foundation and Ocean Futures Society gradually backed out of the Keiko Project.

An article titled "Keiko's keepers plan next move as funds dry up," by Katy Muldoon, *The Oregonian*, Friday, March 15, 2002 stated:

   …(Charles) Vinick, executive vice president of Ocean Futures, which cares for Keiko, said his organization is not eliminating but is scaling back its research operations and efforts to set the famous orca free after 22 or 23 years in captivity.

…Managers have cut back the whale-care staff over the past year, eliminating some American keepers and increasing the number of Icelanders as it became more obvious that Keiko would not swim to freedom quickly, and perhaps may never do so.

The organization still is considering whether to continue caring for Keiko in Vestmannaejar, or the Westmann Islands, off Iceland's south coast, or to move him to a floating pen in an area more accessible for keepers and, perhaps, visitors.

On April 8, 2002 News.telegraph.com.uk posted the following article "Free at last, as Willy finds a family of orcas":

…Charles Vinick, manager of the Keiko project for the Humane Society, which is leading the reintroduction programme, said, "Keiko is choosing to spend all his time with the wild whale pods, staying near them as they mill about and traveling with them when they travel. Last summer he spent just brief periods with them."

…"Keiko is in charge. He has the option of swimming away from our boats and staying near whales. This year he has surprised us by spending almost all of the last three weeks with wild whales."

Dr. Naomi Rose, the Humane Society's marine mammal scientist, who has recently watched Keiko off Iceland, said: "We're very encouraged by his progress. This is an unprecedented effort to return a long-term captive whale to the wild and it's going very well."

…The aim was to eliminate Keiko's dependence on people. Each summer the whale has made progress after being led out of the bay that is his winter home, and has shown growing interest in the wild whales.

…Paul Irwin, the society's president, said, "Bringing Keiko back to his home waters has been a long process and has taught us a lot about releasing whales from captivity."

"We've accomplished what most people said was impossible and almost everybody agrees that Keiko is becoming more wild almost by the day."

KEIKO LOST? OR FREE AND THRIVING?

Summer 2002

Keiko continued to spend longer periods of time in the ocean and was reported to be adjusting increasingly well to life away from humans. His caretakers said he was bonding with other whales, learning to fish on his own, diving over 100 meters deep, and traveling 100 miles a day with wild orcas.

He was fitted with a VHS radio tag, which made it possible to track him in the ocean within a ten-mile range. Keiko also carried a satellite tag that, twelve hours later, enabled his caretakers to monitor where he had been, how fast he was swimming, and how deep he was diving.

On July 17th Keiko voluntarily returned to his sea pen where he was fed before being led back out to join a pod of whales. He was monitored by his handlers aboard their boat and was last fed on July 23rd or 24th. The last time they saw Keiko up close was July 30th, when a storm forced their boat back to shore.

On August 7, 2002 a *BBC* article titled "Free Willy whale 'thriving'" reported:

> Keiko, the whale who found fame in the hit film Free Willy, is adjusting well to life in the wild after years in captivity, according to scientists.

> Experts say that he is now living with a school of killer whales off the south coast of Iceland.

> "We are very excited and optimistic about Keiko's chances of surviving in the wild," said a spokesman for the Ocean Futures Society which is monitoring the creature's progress.

…"Keiko has become more interested in the ocean and other orcas than human beings. That is a very important factor for him returning to the wild," the Ocean Futures Society explained.

August 8, 2002

I was thoroughly surprised by the sudden media reports that Keiko was free and thriving! What a shock, after all the times Keiko had told me and other communicators he would never leave his people. I immediately contacted Keiko to find out how he was and why he had left. This is exactly how the conversation started:

Hello, Keiko. This is Bonnie. How are you doing?
"If the people won't come to me, I will go to the people."

Thinking that was rather ambitious, I asked, *how will you go to the people if you are in the water?*
"I haven't completely figured it out yet, but staying at my home isn't working either. I am limited on my choices and see this as my best option. If I stay in my pen I receive little or no recognition, and if I don't do something now I may not have other opportunities to make things change. I am happy swimming in the ocean."

How do you feel about the people who take care of you?
"They mean well, but they still do not understand me. I have tried to show them in so many ways what I am about, but they refuse to listen. I feel abandoned by them emotionally. I do wish they would understand me for who I am."

How do you feel about all the people in the world who love you?
"They do not think of me much anymore, but I still have love in my heart for them. I miss them tremendously."

How do you feel about swimming off and living in the ocean and not having contact with the people who take care of you?
"I don't have much contact with them anyway. They have shut me out of their lives and hearts. I do miss the love and attention I used to get from them, but it is all different now."

If you could live anywhere in the world, where would you like to live?
"I am still trying to make a statement. I am trying to bring meaning to my life. I don't want to stay cooped up and ignored. I am too aware for that. I

need stimulation in my life and for my mind. I have been very bored."

Please describe what your life would look like if you could do whatever you want.
"I would be free to express myself to others who are willing to listen. So far, not many are willing to listen."

If you could be where people could visit you, how would you feel now about being confined in a small enclosure after having lived in the ocean?
"I like the way my body feels in the ocean. It would be harder for me physically than before. I need a special place."

Would you rather swim off with your whale friends and be free to live in the ocean?
"If I cannot have the attention of the people, that would be better than rotting in the pen. My mind cannot take the dullness for much longer. I am too bright and intelligent to be ignored by everything in my surroundings. If the people will not listen, I must take care of myself."

Keiko, I want you to know that I am still trying very hard to let people know your true story. It may take a few years before it happens, so until then it is most likely that people will not understand you. If your movie can be made, there may be a big reaction and understanding of you. Is there anything else you would like to say?
"I believe in the light of goodness. That all things happen as they should. Your project will light the way for others. I will do what I need for myself and my sanity. I can do little from where I have been other than to waste away mentally and physically. I hope I can survive in the ocean as they believe."

Keiko, I love you with all my heart. Learn and grow from watching your whale friends. Please be safe and healthy and happy wherever you choose to live.
"Thank you for all you have done for me. You are an inspiration to me and I still believe in you."

Thank you Keiko. I love you!

I wrote the following note about how Keiko felt during that communication:
Keiko knows if he doesn't swim off this summer or next, he will be trapped in the bay pen without human interaction and no outside people. He is

making the best choice for himself. He is not a martyr. Staying in his bay pen isn't working or accomplishing anything, and he is unhappy there.

What Keiko was describing made me think of keeping Albert Einstein in solitary confinement without any freedom to explore new ideas or interact with others—a bright, brilliant mind and no one to share it with.

KEIKO COMMUNICATES WITH MARY J. GETTEN

August 9, 2002

In an effort to verify my communications with Keiko, I requested Mary J. Getten ask him the same questions I had asked the day before. It was agreed that she was free to ask follow-up questions in response to Keiko's answers. We did not share our results until she completed her communication.

How are you doing?
"I am doing well."

I heard you are swimming with other whales. How is that?
"Yes, I feel more comfortable with them now."

Do they accept you?
"No, not really, but they tolerate me. They are not aggressive towards me."

Is it fun to be with the whales?
"Yes, without people I need social interaction, so I am trying to get it with these whales. It's fun to really use my body."

How do you feel about the people who take care of you?
"They are great, but they cannot satisfy my needs. I need much more interaction and stimulation."

How do you feel about all the people in the world who love you?
"I wish I could see them and interact with them. I still want to be with people."

How do you feel about swimming off and living in the ocean and not having contact with the people who take care of you?
"I do not wish to swim free. I want to be with people."

Well, then why are you out swimming with whales now?
"I am bored and in need of stimulation. I will not stay with these whales when they move on. I will come back to my pen."

Why don't you change your mind and go free? Wouldn't that be wonderful and make humans happy?
"I realize it would make many people happy, but it would send the wrong message. They would assume that ALL whales want to be free and don't want to interact with humans. This is wrong. I came here to be with people and I would like to fulfill my purpose. My life with people is not over yet. It is so much more fun and interesting to be with people than this life in the wild."

If you could live anywhere in the world, where would you like to live? Please describe what your life would look like if you could do whatever you want.
"I do love being able to dive deeply and swim far. This would be difficult to give up, but I am willing to do it. Perhaps there is a natural enclosure I could be in with a viewing window on one end where the people and I could see each other. That would be best."

Is there anything else you would like to say?
"Do not interpret my time with the whales as a change of heart. There are few people here to interact with, so I am trying to find some fun and excitement. If there were people at my pen, I would be there. For now I have some companionship and am enjoying using my physical body. My greatest desire is to interact with people. Do not forget me."

NOTE FROM MARY: My general impression was that Keiko is very bored and lacking stimulation. He is spending time with the whales because he is starting to think that they may never bring him back to the people he loves so much. He would rather be with whales than die of boredom in his pen, but given the choice, he would choose to be with people.

By mid-August, Keiko had been swimming off the coast of Iceland for several weeks. Although no one knew for sure if he was feeding himself adequately or had bonded with a pod, he was said to be swimming with wild whales. Some reports stated that Keiko was free and thriving with no contact or help from humans.

An article titled "Keiko's new, Free Life," by Katy Muldoon, *The Oregonian*, Saturday, August 17, 2002, said:

> …Charles Vinick of Ocean Futures Society said it would be premature to say that Keiko will not return to his Icelandic sea pen – or that keepers might not lead him back there for the winter if he appears to need help.

> But, "he's clearly free because he's not in our control. He's truly out with whales."

August 23, 2002

On August 23, 2002 I sent both Mary's and my communications to Jean-Michel Cousteau's personal e-mail address. I am especially glad I did this because it documents that Keiko communicated he would go to people three weeks before he actually did. I did not receive a reply.

An article titled, "Ex-caregivers challenge report on Keiko's health," by Katy Muldoon, *The Oregonian*, Friday, August 23, 2002 reported:

> Last week, current Keiko keepers cried freedom.

> This week, former Keiko keepers cried folly.

> Stirring a clash that has flared tempers and wrenched hearts, eight men and women who cared for the *Free Willy* star in Iceland said that the killer whale may be wandering alone and hungry at sea, and that no one is stepping in to help.

> The group, led by Jeff Foster of Auburn, Wash., former director of research and operations for the Keiko project, sent letters this week to the National Marine Fisheries Service and the U.S. Marine Mammal Commission.

> The letters said that current caretakers haven't seen the whale in three weeks, aren't looking for him, and that they cannot prove he is foraging for food or keeping company with other whales— components considered keys to his survival.

> …The society issued a news release earlier this week saying that the orca star of the 1993 box office smash, "Free Willy," had been swimming at sea for 45 days and appeared headed for Norway.

> …When the humane society staff searches for Keiko next,

perhaps when he is closer to Norway, they plan to look for signs of malnutrition. When cetaceans are losing weight they lose blubber around their blowhole, a condition known as "peanut head."

But former keepers say the more responsible approach would be to call Keiko to a boat and extract a sample of his stomach contents through a flexible tube inserted down his throat. That, they say, would prove whether he is eating or not."

Keiko's travel pattern concerns former keepers, too. In the past two summers, during open-ocean "walks," which they called free-swim training exercises, Keiko sometimes would swim far away not only from wild whales, but also from food sources. They worry he may be wandering just as aimlessly this summer.

"I would love for them to prove me wrong," said Stephen Claussen, who worked with Keiko for nearly six years in Iceland and at the Oregon Coast Aquarium in Newport. "But this is the same scenario – that he just picks a direction and goes."

Claussen and others allege that the Humane Society of the United States operates with a moderate to radical animal-rights philosophy, and that the organization appears to be putting its political agenda ahead of Keiko's welfare.

KEIKO COMMUNICATES WITH TYLER GRAHAM

August 26, 2002
On August 26, 2002, I received a call from a very bright seventeen-year-old boy, Tyler Graham. He was so excited because, he said, the animals were talking to him. He explained how much he loves animals and that he wanted to learn as much as possible about Animal Communication.

He told me about some interesting conversations he had had with animals, and that often the animals started the communication with him. He said one time he was with a friend, who was also "into Animal Communication," when they saw a woman walking her dog. Tyler said, "Watch this, I'm going to ask that dog's name...Her name is 'Babe'." They stopped the woman and asked her dog's name. She said "Babe." Even his friend said, "No way!" Tyler told me, "It almost seems too good to be true!"

I was impressed by Tyler's sincerity, accuracy and overwhelming love for animals. At the time he was being home schooled and didn't watch much television, so I asked what he knew about Keiko. Tyler knew almost nothing about Keiko other than he had seen the movie, *Free Willy*. He did not know about any of my work with Keiko or even that he had been taken to Iceland.

I asked if he would be interested in communicating with Keiko. He said excitedly, "Sure!" Then he paused and asked, "Can I do that?" He had only communicated with animals he had been with physically. I explained that it is also possible to communicate with animals long distance.

Tyler happily agreed to try to communicate with Keiko. I told him that Keiko had been taken to Iceland and was now swimming free somewhere. I was very careful not to tell him very much so that his communications would not be influenced by that information.

Because Tyler knew so little about Keiko, his caretakers and his present situation, I feel his communications especially validate our ability to communicate with animals naturally.

Except in conversations he did on his own, which I recorded and later transcribed, Tyler would ask Keiko questions *(in italics)* prompted by me. The following are excerpts from Tyler's communication with Keiko on August 26, 2002:

Tyler: *Is there anything you would like to say?*
"I'm lonely. There is no one here for me."

Tyler to Bonnie: He asked what my name is.
Bonnie: Tell him.

Tyler: *Tyler Graham.*
Tyler for Keiko: He says to help him and love him. He is lonely.

Tyler: *Are you okay? Are you safe? Healthy?*
T for K: He's okay. He's not happy. There's plenty of space.
"Why do you want to know?"

Tyler: *Because we care and are concerned.*
T for K: He said we don't need to be concerned. He's just lonely. They

took someone away from him.

Tyler: *They took someone away from you?*
T for K: Someone with brown eyes or brown hair.

Bonnie: Could this be his trainer?
T for K: Yeah, that's him. Yes, brown, curly hair.

Tyler: *Are you eating enough to stay healthy?*
T for K: They feed him a lot of fish, but he is lonely.

Tyler: *Who is "they"?*
T for K: There's a woman and a few other men.

Tyler: *Are they with you now?*
T for K: No, not now, but they were with him most of the time.

Tyler: *Is there anything we can do for you or to help you?*
"Let me go live on my own. Let me go. I've been here too long. It's all been nice, but I'm tired."

Tyler: *Is there anything else you would like to say?*
T for K: He wants to thank his trainers for what they have done for him.

Tyler: *How do you feel about being with people versus being free in the ocean?*
"I'm tired. I'm tired. I want to go home."

It was getting late so Tyler and I arranged to continue the conversation the next day. In an attempt to keep Tyler from being influenced, I asked him not to read or watch news reports about Keiko while we were communicating with him together. Tyler agreed.

August 27, 2002
Knowing Tyler was going to communicate with Keiko, I made a point of communicating with him first. I always did this to reduce the possibility of being influenced by the results of other communicators.

Keiko, how are you doing, my friend?
"I am happy to swim in the ocean."

Keiko, no humans have seen you for several weeks. Ocean Futures,

the people who cared for you until recently, are concerned for your health and well being. How are you?
"I am swimming free, yet I have no understanding how to behave."

Are you with other whales?
"No, not at this time."

Are you comfortable being alone?
"Not really. I'd rather be with others who can guide me."

Why are you not with the whales you were with before?
"They really didn't want me to go with them so I had to leave."

Are you eating enough to stay healthy?
"I don't think so. Some of my old fish would taste so good."

Are you hungry?
"Yes, it's hard for me to eat enough."

How do you feel physically?
"My body is tired and sore. Even though I am big and strong, it takes lots of energy to move."

Would you like people to help you in any way?
"Yes. I would like some food. I am hungry."

Do you want to go back to your bay pen where you have lived for the last few years?
"Only if they offer me love and attention. I want my people back. The new people are not my people. My people know who they are. They have given of their hearts."

Keiko, maybe you should return to your bay pen so they can feed you and help you become strong again. After what has happened with you being on your own, maybe they will reconsider your future.
"I cannot tolerate living there again as it has been. I have much to do in this world and cannot accomplish it there."

What do you have to do in this world?
"Bring messages of peace and welfare for all. I cannot stand the isolation. I need people to interact with and influence."

If you stay in the ocean, you will need to swim where it is safe for the winter. Can you do that?
"I am not sure what I am capable of. I only know I don't want to go back to things as they were."

If your true people (the original trainers and caretakers from Ocean Futures) agree to interact with you again, will you come back to the bay to get strong and healthy?
"Yes, I would love to be with them again. They bring meaning to my life when they acknowledge me and love me. I miss the interaction with them."

Keiko, how about if you start swimming slowly back to your bay pen. Pace yourself so you stay healthy. I will let Ocean Futures people know what you have said, and hopefully now they will believe that you want to stay with people and help make that happen.
"What if they don't or can't?"

That would be sad, wouldn't it?
"I don't want to be there by myself anymore."

What do you think you should do?
"I am confused too. I am trying it their way and I have no friends here either. It's like no one wants me, and I use to be loved by so many. It is sad. I am sad."

I will do what I can for you. Please help them locate you.
"Will they come to me?"

I think so. They want to check on you. Is there anything else you would like to say?
"Thank you for your kindness and concern. I wish more people loved the animals as you do."

Thank you, Keiko. Please be safe and healthy and feed yourself.

KEIKO COMMUNICATES WITH TYLER GRAHAM

August 27, 2002
Tyer explained to me that "as long as I ask in my mind, Keiko really acts like I'm talking to a human. I mean, I talk to animals like I talk to people.

They get back really fast—like a conversation."

We continued our communication as I prompted Tyler with more questions to ask Keiko:

Tyler: *Are you alone or are you with other whales?*
"Alone, I've always been alone. Why?"

Tyler: *Why are you not with the whales you were with before?*
"They got annoying. They got territorial and what not."
Tyler to Bonnie: I have no idea where that came from.

Tyler: *How are you feeling physically?*
Tyler for Keiko: He's feeling fine, he just misses his people. His staff. His crew. His people. He's lonely because no one is there to talk to. The whales outcast him because he has big ideas about humans and animals and bringing them together.

"The other whales could care less; they just laugh. They have a very negative attitude. They don't see what I see. They don't see what humans can be—what you can become. What people can do for us animals. They don't see that. The whales here don't understand, they don't accept me, and they don't see my vision for my friends, the humans and animals. Bonnie, you will be my voice to the world. You will be my light. And then you can help."

Tyler to Bonnie: He said I can help!
Bonnie: That's what I'm thinking. Tyler, please tell him thank you very much.

Tyler: *How do you feel about going back to your bay pen?*
"I'd love it! I would love it. All my friends are there. I am here."

Tyler: *What friends?*
"The people who smile at me. 'Joe.' A light in their eyes. I miss that very much and the attention I used to get. Down here I don't get so much as a smile."

Tyler: *Can you find your way back?*
"No. No, I'm lost. I miss it so desperately. I don't know where to start."
T for K: He can't find the bay pen.

Tyler: *Can you tell us anything else to prove this (animal communication) to your people?*
"It's not that hard to believe. It's like echolocation. Humans, you people are so stubborn."

August 28, 2002
I checked in with Keiko the next day.

Hello, Keiko. Is there anything you would like to say?
"I am confused. I'm not sure what's expected of me. If I can only find my way back to my people."

Are you saying you're lost?
"Yes. I am disoriented."

Are you alone or with other whales?
"I am alone and not sure where to go."

Are you eating enough to stay healthy?
"My body is slowing down. I am tired. I would like to go home now."

Where is home?
"With my people, of course."

Are you feeding yourself?
"Not really enough for the demands of my body."

Can you find your way back to your sea pen?
"I think so, but I am very tired."

How can your people help you?
"Please find me and bring me home. I miss them and I am very tired. Look for me in their boat. I know their sounds. I will be looking for them. This is all too overwhelming for me. I need their help."

How are you feeling physically? (I sensed Keiko floating very still. I deliberately imagined him resting because I did not want to consider he had left his body. He needs our physical and emotional support to help him swim back. He needs our encouragement, thought and caring!)

Thank you for communicating. Is there anything you would like to say?
"I have been very traumatized. I want to go home. I need the help of my

people. Please let them know this and help them to find me."

Thank you, Keiko. I will tell them. Please be safe and slowly head back to your bay pen home. I love you Keiko, with all my heart.

KEIKO COMMUNICATES WITH MARY J. GETTEN

August 29, 2002
Out of concern, I again contacted professional Animal Communicators, Mary J. Getten and Teresa Wagner. I requested they ask Keiko questions based on the ones I had asked him during the last couple of days.

Hello Keiko. Is there anything you would like to say?
Mary for Keiko: I feel Keiko logging (floating) on the surface. Waves are slapping gently along the side of his head.

How are you feeling physically? Emotionally?
"I don't have much energy. I am very tired and sad. I am lost and I want to go home, but I don't know where that is."
M for K: He feels very depressed physically and emotionally.

Are you alone or with other whales?
"I am alone. I haven't seen other orcas for almost two weeks. I am far out—there is no land in sight."

Are you healthy?
"I am not healthy. My energy is very low."

Are you eating enough to stay healthy?
"No, there is not enough to eat. I am very hungry."

Winter is coming. What are your plans for the near future? If you want to return to your bay pen, can you find your way back? If not, what can be done to help your people find you?
"I want to go home to my sea pen, but I don't know how to get there."

Do you in any way need help from your people?
"Yes, I DO need help. Have them come and find me. Tell them to bring food. I need food."

Describe where and how you would like to live for the rest of your life.

"I would like to be in an enclosure with a window where I can see people. I must be surrounded by people. I want to interact with people."

Is there anything you would like to say to your people from Ocean Futures?
"Please come get me. I need help. I am very sad and alone. Come get me."

NOTE FROM MARY: Keiko felt to me like he was very depressed and had possibly given up. He wasn't trying to fish or go anywhere. He is hopelessly lost.

Keiko Communicates With Teresa Wagner

August 29, 2002
Hello, Keiko. Is there anything you would like to say?
"I am very tired, and I'm kind of lost. I've been out here because it's what they seem to want of me, but I'm not good at being alone—physically or emotionally."

How are you feeling physically? Emotionally?
"I'm not skilled enough to be out here alone—it's so big."
Teresa for Keiko: He actually used the word "vulnerable" at times, but I wasn't picking up abject fear; just fatigue and tiredness and being lost.

Are you alone or with other whales?
T for K: He said he is alone. He is not with other whales.

Are you healthy?
"Not really."
T for K: He needs external help. He said, "I don't know how to self-heal, I only know how to help others heal." The good thing is he wasn't feeling fear or terror; he's just uncomfortable.

Are you eating enough to stay healthy?
T for K: No, he's not eating. He's not eating at all.

Winter is coming. What are your plans for the near future? If you want to return to your bay pen, can you find your way back? If not, what can be done to help your people find you?
T for K: He's not even in a position to make plans. But what he wants is

to go back to his bay pen. He definitely wants that. But he says, "I am vulnerable. I have no idea how to get there. I need help."

Do you in any way need help from your people?
"Yes, please come get me. Please come get me. I'll know you when you come. Please bring fish!"

Describe where and how you would like to live for the rest of your life.
"Oh, near the children. Please let me be near the children. Bring the children back to me!"

Keiko Communicates With Tyler Graham

August 29, 2002
Tyler called me the night of August 29th. He told me he had communicated with Keiko on his own and wanted to share what he said.

Tyler for Keiko: Keiko is very deep. It's more of an inspiration. He wants to be heard. He misses home, where people are. He loves being around people. And, he actually misses being around them. He said he is too important to go back to his pen. And, it is too important not to be heard. He wants peace between humans and animals.

Bonnie: Do you know specifically what he wants to have heard?
T for K: That animals are something more than what they are. Like, they are wonderful companions, but he wants everybody to know animals are more. Without animals, humans would have a lot of trouble. Animals help us so much more than we know. And, it's not just physically, you know when you come home and your dog greets you at the door. They're forgiving. Animals give us inspiration. It's more that they give you thoughts; they give you confidence, just by looking at them.

Bonnie: Did Keiko say anything else to you?
T for K: It's just that he wants to be heard. He thanks people like you a lot. And, maybe me later on when I start doing things. He wants to thank Animal Communicators. And you could like, tell everybody. It's hard to explain. You could kind of be his voice to the world, and help everybody understand animals are more.

Bonnie: That's exactly what I'm trying to do.

T for K: And, he said when he is with killer whales, he gets along with them, but they don't accept him as much. I think it's hard for him because he wants to be around people. He's too important to go back to sitting in a little pen by himself. He's amazing!

Bonnie: Did you get anything about how he's feeling?
T for K: He's feeling okay. I think he really misses home. He misses back where all the people are. They bring him such great gifts. We can't just put him in a pen and leave him there. He doesn't want that. Keiko is an incredible animal. He's too important to be locked away. I almost started crying when I was talking to him.

Bonnie: Tyler, what is so great about you letting me record you with all your enthusiasm, is that you're proof that we are born with this ability and most of us just shut it down. Nobody told you anything; you're just getting it. And, you're getting the same information we are.
Tyler to Bonnie: I love, I love animals. And, I believe this stuff. I love animals. I can just sit with them all day.

Bonnie: When you communicate with him, try to get real quiet and ask him to be as clear and accurate as possible, and always thank him. I think it's important to ask him questions about how he's doing and if he's eating, and that kind of stuff.
Tyler: Right. Can I ask him like, "What kind of advice do you have for people?" Like, people as a whole?

Bonnie: You can ask him whatever you want.
Tyler: Sometimes, is it normal to get unclear stuff?

Bonnie: You can ask him questions like, "I didn't quite understand. Could you please show or tell me in a different way?"
Tyler: Right. Sometimes it's foggy though.

Bonnie: You can clarify it with him too—just like you would talk to a person by saying, "So what you meant was, you're too important not to be heard?" And he can verify it by saying, "Yes," or "No."
Tyler: Right. When you get information, is it fast?

Bonnie: We all get it differently. Many Animal Communicators get it just like you. Then, sometimes they have to sort of translate or interpret it a little bit. Don't try to do it like me or anyone else. Do it your way, trust it,

and verify it with him.
Tyler: Thank you so much.

August 31, 2002
Hello Keiko, my sweet one. Is there anything you would like to say?
"I can barely hear you."

Keiko, I would like to communicate with you and know how you are doing.
(I then thought about reports that he is diving deep in the ocean.)
"No, I am not diving. I am here on the surface. I do not have the energy to swim, much less dive. This has all been very hard on me. Why won't people listen to us?"

Keiko, I'm afraid some people have their own agenda and that doesn't include what is best for the animals. How are you feeling physically?
"I don't even feel much anymore. I am weak and tired. I want to go home."

Keiko, thank you for giving so much love to the world. People who have seen you will never forget not only your magnificent presence, but also the way you touched their hearts. When people came to see you, and you would look at them—it was almost impossible not to feel the love you would send to everyone. You are a most incredible being, my beautiful one. Please know that you helped open their hearts to the animals and to themselves. You gave so much of yourself Keiko, it's hard to thank you enough. Keiko, there may be a boat on its way to help you. Can you hold on until it arrives?
"I am dizzy now. My mind and body are softening."

What do you mean?
"I'm not sure how much longer I can hold on. I am very weak. I'm not sure I even want to."

Keiko, as much as I want you to be found and made healthy again, I think I understand.
"What point is there in struggling to get healthy again if my life is full of dullness and no purpose?"

The only thing is, maybe people will listen to you now, but they still may not be able to provide the life you desire. This world can be a sad place.
"I know. I have tried to help by sharing my heart with others, and am grateful for the opportunities I have had."

He showed me a picture of him in his small tank in Mexico.
"Even doing tricks touched people's hearts in ways that otherwise would not. I am indeed grateful for my life here and what I have been able to accomplish. Please thank those involved with that part of my life and all those who have cared for me over the years. I connected with them in ways they will always remember."

Keiko, you are a most beautiful soul. I am grateful that I was able to see you and be able to communicate with you. You have enriched my life and opened my heart. When I first saw you, I fell in love with you. I could feel your love and kindness every time I saw you and with every communication. You have been an inspiration to me and I love you with all my heart. Is there anything else you would like to say?
"Bless the people of the planet. Most of them mean well. This can be such a beautiful place if only humans would learn to love and care about each other. We cannot go on the path we are on. There is much for man to learn."

Thank you, Keiko. Please be strong and hold on for your people to find you.
"I will try. I too, am disappointed by humanity."

Keiko, I am going to tell the world your story! People will learn of you and your life and all you gave. Your gift of giving and sharing love will go on and affect many, many more people. I will never give up on you and what you have given to the world. I will continue to keep your legacy alive.
"Tell the people to go inside themselves. To look at their heart. To ask themselves what is their purpose in life and are they honoring that? Listen to your heart and it will guide you."

Thank you, Keiko. Is there anything else you would like to say, my friend?
"Continue on your path—it is a good one. Help the animals. Show people what we are about. You are a link between the species."

Can you explain what you mean by showing people "what we are about"?
"You know. Our purpose is to teach people of love and kindness, and to believe in us and themselves. There is no difference in species. At heart, we are all the same."

Keiko Swims To People In Norway

September 2, 2002

After swimming over 800 miles, Keiko surprised the world by arriving in a Norwegian fjord where children got into the water and played with him!

It didn't take long for the family in their boat to identify Keiko, especially once they saw the satellite tags on his dorsal fin.

The Associated Press, September 2, 2002 described the event in an article titled "Keiko the whale frolicking in fjords":

> …The orca surprised and delighted Norwegians, who petted, swam with him and even climbed on his back in the Skaalvik Fjord, about 400 kilometers (250 miles) northwest of the capital, Oslo.

> "He is completely tame, and he clearly wants company," said Arild Birger Neshaug, 35. Neshaug said he had been out in a small rowboat with his 12-year-old daughter, Hanne, and some friends when they spotted Keiko on Sunday.

> "We were afraid," Neshaug said. "But then he followed us to our cabin dock. At first we were skeptical, and then we tried petting his back. Finally the children went swimming with him." He said the orca stayed by their dock all night and into the day on Monday, happily eating fish tossed to him by the families.

> Ugarte is monitoring the whale on behalf of the Ocean Futures Society and the Humane Society of the United States. He said Keiko was in excellent shape, but still seems to prefer humans to other whales.

Keiko: "I am happy now. I have found children again!"

Photo: Retna

It had been six weeks since he left his bay pen in Iceland, and ironically, Keiko followed a Norwegian fishing vessel into the only country that still hunts whales commercially.

There was some debate as to whether Keiko had actually fed himself during the six weeks he was on his own. From all outward appearances he looked healthy, but I later learned that it would have been possible for him to go as long as two to three months without eating before a significant weight loss would be noticed.

On September 2, 2002, *KOMO* in Seattle, Washington reported "Keiko doing well in Norway":

> ...Nearly 60 days after the orca whale Keiko left his sea pen in Iceland this summer, visual observations made in Norwegian waters confirm that he is in excellent health.

> "I have reviewed the photos just taken of Keiko, and it is clear to me that Keiko is fit and thriving," said Dr. Lanny Cornell, Keiko's lead veterinarian. "After 60 days at sea and traveling more than

one thousand miles, Keiko is strong and does not appear to have lost any weight whatsoever..."

Keiko's former trainers, Jeff Foster and Stephen Claussen later submitted the following joint statement to me regarding their view of Keiko's status at that time:

When Keiko left Iceland after being abandoned by his then caretakers out at sea, we predicted that he would follow a boat into a port somewhere looking for human interaction and, more importantly, someone to feed him. Just a couple of weeks later, he did just that.

Keiko arrived in Norway where he was documented to be lethargic, but who wouldn't be after being lost and not eating for so long? He exhibited solicitation behaviors, ("gaping" or opening his mouth while lying on his side, looking at people), that we had seen before when he wanted either companionship and/or food. The organizations that took over Keiko's care after our team left were saying he had fed himself the entire time but was tired from his journey.

We were very concerned for Keiko and offered to fly to Norway, paying our own expenses, to check his overall health, offer any help needed, and provide real scientific documentation as to whether he was feeding himself or not via a stomach sample.

Keiko had been taught how to receive a stomach tube orally a long time ago. It was one of his strongest, (best performed), behaviors and would have been spectacularly easy. It would have conclusively proven whether or not Keiko had been able to feed himself in the wild. Scientifically documenting this fact was very important in our professional opinions because the Keiko project was an experiment never before attempted. They should have been documenting and learning as much as they possibly could every step of the way, as our team had done from Newport, Oregon to Keiko's second reintroduction season in Iceland before we left the project.

The organizations then in charge of Keiko's well being denied us permission to come and help this amazing animal and further learn from his incredible adventure.

We always felt that what was in Keiko's best interest was what should always have guided the project. The whale and the project should have been handled more responsibly—that was all we ever asked for and Keiko deserved it. In the end, it was an opportunity to learn so much, yet this golden opportunity was allowed to slip away in the end.

At the time, Mary and I were anxious to find out how Keiko was doing and why he had been communicating he was lost, hungry and not feeling well. Teresa was unable to participate in our communication this time. Again, Mary and I asked similar questions and did not disclose what Keiko said until we had both finished our communications.

The following is my communication with Keiko on September 2, 2002.

Hello, Keiko. Is there anything you would like to say?
"I am happy now. I have found children again! I touched their hearts and the hearts of many around the world."

Yes, Keiko you are incredible! And, what a wonderful way to show the world how much you want to be with people. I think that will help us prove what you have been saying all along. How did you find your way to them?
"My heart found them. I followed my heart."

Keiko, how are you feeling physically?
"My heart is full of love again. For those moments, it is all worthwhile. I can even see myself with the hope of being with people again. It is so important to me."

Keiko, I am so proud of you! You are magnificent! I would also like to ask you why so many people picked up that you were lost, hungry and not feeling well physically? And, why did you tell me and several other Animal Communicators that as well? Can you explain?
"Yes. It was 'A call for attention.' I needed to have everyone's attention or they would not pay attention to what is really happening. That was the only way to have so many listen and be aware of me. Then with the children—now everything has more meaning than it would have otherwise. If I had just swum up to the children, it would not have been as noticed. I have many more people behind me and my purpose now. People who care about me got concerned. They got involved. They are

more willing now to help me than they would have been otherwise. Do you understand?"

Yes, Keiko. That does make sense.
"Now, you can take those who know I don't belong in the sea and together help me to be with people. While I was out, people were concerned. People thought about me for the first time in a long while. I took their thoughts, their energy and tried to make a statement. Between what you communicate with me and what I was able to show, gives us much more to help others understand how I want my life to be. I want to be with people. Everyone needs to understand that."

Thank you for explaining that. Is there anything else you would like to say?
"Thank the others for being messengers too. You have much work ahead of you. Thank all the others too, and remind them to continue thinking of me. We must not lose the momentum we have just created. Love to the World, Keiko."

Thank you, Keiko. I am so happy you are all right! I will continue to work even harder to help you again be with people. I love you with all my heart.

KEIKO COMMUNICATES WITH MARY J. GETTEN

September 3, 2002
Is there anything you'd like to say?
"I am so happy here. This is what I want, what I have been missing—the love of the people. I have been so lonely and sad in my sea pen with no interaction. Now I have finally found people who love me and want to play with me again. It makes me so happy."

How are you physically?
"I am fine. I could be eating more, but overall my health is good."

Why did so many people get that you were hungry? Can you explain?
"I was hungry for attention and love and I sent this message overwhelmingly to everyone who would listen. My spirit was dying. I did not know that I would find people who would love and care for me and whom I could love as well. I was just swimming with no land in sight. When you and the others reached out to me, I knew I had to make a strong statement to get help. And so I did. I had given up on my

life. Emotionally, I was devastated and I sent this so strongly that everyone thought I was actually physically dying. I had to get your attention."

Well, Keiko, you certainly got our attention. Did this help you?
"The love that I felt coming from everyone kept me going. Once I felt that, I knew that if I just kept going I would find more people. It gave me hope and let me know that I am cared for."

What do you want now, Keiko?
"I want to stay with people. I do not want to be free in the ocean with whales. I need to be with people and they need me. Why can't they see this? I have been saying this for all these years and no one will listen? I MUST be with people. They are my family. Please tell them this."

You really scared me, Keiko. I believed that you were actually in real trouble. Do you realize that this might make people disbelieve what you tell us? We told them that you were sick and now we find out that you're not. It doesn't engender much trust in what we tell them.
"I am very sorry. This was the only way I knew to get your attention. My spirit was dying and I had to get help. Many people still care about me, and I want so badly to be with the people. Please make them understand."

I suspect that the Humane Society will try to lead you back to your sea pen and then isolate you there again for the winter. How do you feel about that?
"I do not want to be alone in the sea pen. If there are people there who will play with me and interact, that would be fine. But to just be left alone is a terrible way to treat someone. I will die of loneliness."

I think they are isolating you so that you will become less dependent on people and want to make friends with whales. They are trying to help you become a free whale again.
"This will never happen. I was made dependent many years ago and now my heart is full of love for everyone."

I hope that they will understand, Keiko, but scientists do not see things as we do.
"I will find my way back to people again if they release me. Have them take me to an enclosure where I have a natural environment, but people can come and see me, feed me, and play with me."

"Why can't they understand this? I am Keiko, the captive whale. I may survive physically, but not emotionally, in the sea. You can see this by how happy I am to have people to interact with again. This is my desire."

KEIKO COMMUNICATES WITH TYLER GRAHAM

September 3, 2002
People in the area immediately flocked to see Keiko. Those responsible for him began urging visitors to stay away. They were concerned that contact with people would set back Keiko's reintroduction to the wild. Although I didn't say anything to Tyler, steps were already being taken to keep people away from Keiko.

Tyler: *Keiko, how are you feeling physically?*
"Not very good. I'm feeling lonely and think I'm starting to get sick."

Tyler: *Are you eating enough to keep your body healthy?*
"No, the food source is really low over here."

Tyler: *Do you need any help from humans?*
"Any help I can get from humans is wonderful, but if they won't come to see me, I will come to them."

Tyler: *How can your people help you feel better?*
"There's nothing you can really do except wait. But what I'd really like is to have some people over here—I'd love to see their smiling faces again."

Tyler: *Does not seeing people have anything to do with how you're feeling?*
"It's the biggest part of my spiritual lacking right now. A piece of me is gone right now."

Bonnie: What does he mean, "a piece of me is gone right now?"
Tyler to Bonnie: To give you an example, have you ever loved someone and then you lost them? Never be able to see them again? It's like a piece of you is missing. It becomes so common over your life. And once they're gone, you become physically ill.

Tyler: *What do you want right now?*

"What I want is not what the world wants. People want to set me free. They think it's the best for me. I know they're trying to do good, but they are not doing what I want. I want to see the crowds and their smiling faces. I want to see the kids come out and put their hand on the glass. I want to be free. I want to be progressive. I'm in a lack of progression. I learn so much from humans and I know they learn from me."

Bonnie: What did he say about being free?
"Free with my brothers, free with my sisters, free with my people."

Tyler: *Your new people will probably be coming with a boat for you to follow back to your bay pen. How do you feel about that?*
"That would be absolutely wonderful!"

Bonnie: Ask him why he told me he didn't want to stay there anymore. How does he feel about staying in the bay pen all winter again?
"I will only go back if the people who support me and love me are there."

Bonnie: If he goes back to the sea pen, I believe it will be the same as it was before.
"I have such big plans for the world. I need everybody's love and support."

Tyler: *How do you feel about going back to your bay pen if they are going to treat you the same way?*
"Absolutely not. I will not stand for that at all. I am too important to just sit there with no one there. I need thoughts. I need human emotions. I crave that. They have no idea. They think they know what I want, but they don't know. They just know what they want."

Tyler: *They may bring a boat for you to follow back to the bay pen. What do you want to do?*
"Yes, I will follow the boat back. Any human companions are better than none. But, I will not be ignored."

Tyler: *How do you plan on doing that, when it is up to your people to interact with you?*
"Bonnie will explain to them about me."

Bonnie: Yes, I will. But they will not listen to me.
"Bonnie will explain my plan about my life. Bonnie is my voice to the world."

Bonnie: Winter is coming and, even if I were able to convince his people, it will take time. Ask him what he wants based on this information.
Tyler for Keiko: He wants to come back to the pen, but he really needs human interaction. I told him it will take you time to talk with his people.

Bonnie: Tell him there's no guarantee. I'm up against the Humane Society of the United States.
Tyler: Really?

Bonnie: It's not in my control. I can try, but that's all I can do.
"If you can't set up that meeting or convince them, then you leave that up to me."

Bonnie: Leave that up to him?
"Right."

Bonnie: Well, he certainly had the most fantastic plan that anyone could ever have come up with—what he did yesterday!
"Thank you. I thought of it myself."

Tyler: *Are there people around?*
"No, if there were I would go see them."

Tyler: *Are there people there that want to see you?*
"They want to, but I don't think they can."

Tyler: *Why would that be?*
"I think you call them the police—won't let them."

Tyler: *Is there anything else you would like to say about that?*
"I won't hurt them. I love the people. I don't know if they're afraid of me or what?"

Tyler: *Your new people are keeping the public from getting close to you because they don't want you to be around people.*
"Why would they do that? That's awful."

Tyler: *Because they believe you should be out there free, not connected to people.*
"I understand that they don't know what I want, but sometimes humans can be so selfish."

Bonnie: I have tried to tell them many times, but they won't listen to me. Does he have any suggestions?
"Like I said, if the people won't come to see me, I will come to see the people."

Tyler: *And, how will you do that?*
"I will go on land if I have to."

Bonnie: No, please don't do that! Tell him, please stay in the water! You must stay in the water to stay healthy.
"I was just kidding anyway."

Bonnie to Tyler: Do you know that whales and dolphins, along with humans, are among the few mammals that are capable of committing suicide?
Tyler: No, I didn't know that.

Tyler: *Keiko, are you saying you will keep coming back to land after they turn you loose?*
T for K: He will keep coming back to land.

Tyler: *Is your plan to keep coming back to shore?*
"Yes, there are people there."

Bonnie: Ask him again just to clarify, why did he tell everyone that he was lonely and lost and hungry? Why did so many people sense that he was not doing okay, and why did he give us that information if it was not correct?
"Wouldn't you too if you were separated from your family?"

Bonnie: How does that explain being hungry and sick? He showed up and he looks like he's in pretty good shape.
"Like I said, hungry for companionship."

Tyler: *What about being lost?*
"I am lost. I am searching for my people. I am lost."

Tyler: *Who are your people?*
T for K: Mexico? Joe?

Tyler: Is that right? I got the name "Joe."
Bonnie: I don't know. His trainer's name was Jeff.

Tyler: *Is there anything else you would like to say?*
"Yes, there is. I love humans. I love the spirits. I love the heart; even the most disagreeing, I love. And even if it takes me a lifetime, I will get the respect I deserve, and we will share it together. Thank you so much for all of your help."

Tyler: *Thank you. It's been wonderful talking to you.*
"You know, as spirit beings we are all in search of acceptance and love—everybody and the animals are in search of acceptance."
T for K: He also says we all want to be a part of something bigger than just ourselves.

KEIKO COMMUNICATES WITH TYLER GRAHAM

September 4, 2002
On Thursday, September 5th, I received a call from Tyler saying Keiko had communicated with him the night before. Tyler was getting ready to go to bed and, as sometimes happens, he had not made any effort to communicate with Keiko. Instead, Keiko contacted him. Tyler said, "It was so strong!" He wrote down their entire conversation and read it to me over the phone.

Keiko: "Tyler, I need you."
Tyler to Bonnie: Like he sought me out.

Tyler: *Do you know who this is?*
"Yes, it's Tyler."

How have you been feeling lately?
"I've been okay, but I miss my people dearly."

Have you visited anyone lately?
"I wish I could, but no one's around. I'm debating whether to go on land."

Keiko, please stay in the water!
"That's only a last resort."

Keiko, we love you and we want the best for you.
"Thank you, and I appreciate everything you are doing for me. One day me and my people will be reunited again."

Keiko, what's your goal in life?

"My goal is for humans and animals to come together so we can pull strength from each other and put aside our differences and help each other achieve spiritual happiness."

Thank you. That was inspiring.
"You're welcome."

Keiko, how are you feeling physically?
"I'm doing okay physically. I feel as though I am getting sick though."

Is there anything your people can do for you?
"Just come together so we can pull faith from each other."

I'm sorry Keiko, I need to go to sleep now. Thank you so much for contacting me. I always enjoy talking to you. I love you.

KEIKO COMMUNICATES WITH SUSAN GATES

September 4, 2002
What's even more amazing is that later that same day Thursday, September 5th, I received a call from Susan Gates. Susan had already been communicating with animals when she attended a workshop I had taught earlier that year. Based on experiences she shared, Susan was obviously very intuitive and communicating accurately.

She was crying because Keiko had also connected with her the day before and told her he was thinking about beaching himself! I told her about Tyler's conversation and encouraged her to join us by communicating with Keiko as often as possible, and asking him to please "stay in the water." Susan also said when she was thinking about Keiko she imagined just putting her "arms around him and hugging him."

September 5, 2002
Because of the seriousness of this, I asked Penelope Smith, known worldwide for her work in the field of Interspecies Communication, to see what information she might receive from Keiko.

Penelope replied, "Keiko confirms the information that you got about him thinking about dying and wanting to be with people. He wants our help to bring him 'home' with people."

On September 5, 2002, *The Associated Press* described Keiko's situation in an article titled "Keepers fear Keiko the killer whale and movie star is victim of his own fame":

> It's a misty Thursday morning and Keiko the killer whale is alone, a rare sight since Norwegians discovered the 'Free Willy' movie star in a fjord and flocked to pet and play with him."

> …But six weeks after he was released from his pen in Iceland, his keepers are worried that Keiko is becoming a victim of his own celebrity and are begging people to leave him alone, so that the $20 million project to return him to the wild can continue.

> "We are just trying to get people to leave Keiko alone so he can decide whether he wants to stay here," says Colin Baird, a 35-year-old animal trainer from Canada. "All the distractions from boats and people are keeping him from doing what he should by hunting food."

> Keiko seems content in the fjord in western Norway, where he ended up last weekend after swimming 870 miles in what many said was a search for human companionship.

> On Thursday, he swam under an empty, moored yellow boat and used the barnacles on the bottom to scratch his back, then his white belly.

> He surfaced, exhaled from his blowhole and peered around as if to check whether his fans had arrived. He let out a cry and was answered only by the bellow of a nearby cow.

> The previous evening, Keiko was overrun by fans, including 16-year-old Mona Lindkvist.

> "It was silly to set him free. He wants to be with people," said Lindkvist, who rowed for hours in a small boat to pet him near the floating dock at the Skaalvik Fjord in the township of Halsa.

> …Lindkvist was joined by about 70 people standing on the flimsy floating dock. Keiko lay placidly in the water, surrounded by a dozen boats. Obligingly, he edged closer to the dock and seemed to wave with a fin, occasionally "talking" to people in chirps.

> A father on a boat held his daughter by the belt as she hung upside-down trying to reach Keiko. Some whistled or did their best to imitate whale calls."

"It would be wonderful if he stayed here," 13-year-old Tanja Kristin Haugen said.

…As Norwegians ponder what to do about Keiko, Baird and his colleague Fernando Ugarte, a marine biologist from Mexico City, tour local schools and explain why Keiko must be left alone.

At the Blakken Primary School, Ugarte, who speaks Norwegian, asked who among the 31 children had seen Keiko. They all raised their hands.

…Norway and the groups behind the campaign to free Keiko, the Ocean Future Society and the Humane Society of the United States, have discussed leading the whale back to Iceland, Baird said. Some have suggested leaving Keiko in a remote fjord and feeding him there."

…Looking tired, Baird sighed. "The amount of human contact has been a bit of a setback. It is a bit frustrating when people don't listen," he said.

Keiko Communicates With Tyler Graham

September 5, 2002
From the time Keiko arrived in Norway, I was very busy gathering information, contacting people, writing press releases and a letter to the organizations responsible for Keiko. I greatly appreciated Tyler's willingness to communicate with Keiko during this time. Tyler asked Keiko the following questions prompted by me.

Tyler: *How are you?*
"I'm okay."

Tyler: *How are you feeling physically?*
"I'm okay, but I miss my people desperately. I mourn for them."

Tyler: *How do you feel about what you said yesterday about going on land?*
"If people don't come to me, I'm going to have to do something drastic to attract the attention of the people."

Tyler: *Do you understand if you do that, you will get the attention, but you will die?*

"People will know I did something good for the spirit."

Tyler: *Keiko, please listen carefully. This is very, very important!*
"Okay."

Tyler: *Bonnie is working very hard to unite those who want to support you in living with people. She thinks she may be able to pull it off so you can live in a place with people.*
"That's wonderful news. That makes me want to cry."

Tyler: *It is very important that you stay in the water. It's definitely going to take some time for this to happen, but the world is behind what you want.*
"I will be patient."

Bonnie: Tell him he's been so patient; bless his heart!
"It's okay, I understand everything you are trying to do for me."

Tyler: *Bonnie is feeling a lot more optimistic than even a few days ago.*
"That's wonderful."

Tyler: *Is there anything else you would like to say?*
"Let the people know that I'm there for them just like they're there for me."

Tyler: *Please stay strong and healthy and safe and in the water.*
"Can do."

Bonnie: Tyler, please ask him to contact you and/or me when he needs us.
"Thank you for being there."

Bonnie: Tyler, please check in with Keiko as often as possible so we don't miss a time he really needs help.
Tyler to Bonnie: I can feel him all the time—while I'm working, at school, at Tai Kwon Do. He talks to me. Everything is open. My dog talks to me when I'm doing homework. Sometimes for me, it's not getting quiet, it's just my love for the animals and acknowledging what they're saying.

Tyler then shared an experience he had visiting a pet shop:

It's really hard though, because all the animals have so many things going on. The rats felt really bad because everybody just passed by them. People don't really pay attention to them. They said, "Hey, down here! Down here, we never get attention. We're just as important as everybody else here. We don't understand why people don't love us. They don't appreciate us for who we are. You can talk to us. You, you listen."

The birds said, "Don't look at them, look at us. We're beautiful, look at us. We're colorful and bright. We can make you happy!"

Then the hamsters said,
"Come on, we're the cutest. We're the best bargain in the store."

It was really hard. I asked this lady, "What about the rats?" She said, "Whew, no!"

I'm like, oh, boy. This is hard. Almost everybody wanted to come home with me, and I can't do anything.

I didn't promise anything, but told the rat, *I'll ask my mom if I can take you home. Would you like to come home with me?*
"Sure! Are you going to let me out?"

I'm going to let you out so you can roam around the store, but you have to promise me you won't run away.
"I won't. I promise."

Then you'll have to go back, okay?
"Okay. "

I put him on the floor, and he kept coming back. Then he let me put him back in his cage.

Bonnie: So, you did bring him home?
Tyler: No. I asked my mom, but she wouldn't let me have him.

Bonnie to Tyler: I encourage you to keep talking with the animals so they can be heard, and send them love and positive energy and thoughts for them to all go to good homes.

September 6, 2002
Hello Keiko. How are you, my friend?
"I am doing well."

How are you physically?
"It is time to feed me."

Your people are not feeding you so that you will swim back to the ocean to live.
"It will never happen. I would never volunteer for such a life. I won't even do it if it is forced on me."

Keiko, are there people near you?
"I try to see them."

Keiko, I want you to know that I am working very hard to unite the millions of people who want you to be with people. Please be strong emotionally and physically—this will take some time.
"Thank you."

Please be patient. Thank you for your incredible patience so far. Just a little longer. Please stay safe and healthy and stay in the water. We can do it, Keiko. You have just provided the way.

On September 6, 2002, *The Associated Press* reported:

> Norwegian officials on Friday barred people from getting near Keiko the killer whale, hoping to protect the star of the "Free Willy" movies from hordes of fans he appears to enjoy.

> Experts leading the $20 million project have pleaded with locals and tourists around the narrow inlet to leave the whale alone, even though he appears to thrive on human contact.

Keiko Communicates With Tyler Graham

September 6, 2002
Tyler had communicated with Keiko on his own on September 6th. Again, I recorded as he read from his notes:

Hello Keiko. Do you want to speak with me?
"Yes, I do. It's not urgent. I'm just lonely, that's all."

Why are you lonely?
"Because there's no one here. I miss the smiling faces of the children and the people's warm spirit. I miss learning from them, just as they learn from me."

I'm sorry. We love you. We're trying the best we can to help you.
"I understand and I thank you. I just need to be patient."

Are you eating enough to stay healthy?
"Food is sometimes hard to get, but I manage. It would be a lot easier if people were around."

Your people love you. How do you feel about them?
"I love them dearly with all my heart and thank them for all the support they have given me. I want to show the world my vision of animals and people—how they can learn from each other. Learn to help each other through easy and hard, and learn to support each other. I know we can learn so much from one another if we just listen. We need to be humble and peaceful."

Thank you for being the amazing, wonderful spirit you are.
"Thank you for listening."

I'll talk to you later.
"All right, my friend. Good-bye."

September 7, 2002

I woke up at 3:45 a.m. with (what I noted as) "A hit to my chest or heart!" I immediately communicated with Keiko.

Keiko, this is Bonnie. Please be safe and healthy and stay in the water. I love you Keiko with all my heart. How are you?
"I am feeling better now. It is so heart warming when people think about me and love me. I am so starved for attention and recognition."

Keiko I am working very, very hard to help you find a place to live where you can be with people. Many others know this is what you want now. You have done the most powerful thing possible to show everyone what you want. It is going to take some time, though. What do you think?
"I am proud to give for the animals. It was my job to touch the hearts of people to help them realize how valuable and meaningful animals are in their lives. I have tried to show people how to be, even in face of opposition. Many will learn from knowing me."

Keiko, many, many more will learn if we can get you to a place where people can come to see you. Can you hang in there for a while physically and emotionally?
"I think so. I have come so far; I won't give up now. I have my life's work in front of me."

Thank you, Keiko. Again, please be strong and safe and healthy and stay in the water, my friend.
"You give me hope and encouragement."

Keiko, know that many others have come to see you already! It may be that you can't be with them right now, but they have come to see you and they love you too.
"It is beautiful for people to make the effort. I will touch their hearts too."

Is there anything else you would like to say?
"My heart is full of gratitude for those who care and respect my wishes. They are the ones who will be changed forever. They will be more heart-driven because of their love. Our hearts have touched."

The following article, "New restrictions are respected by Keiko admirers" by Rolleiv Solholm for the *Norway Post*, September 7, 2002 said:

> The local Association for the Protection of Animals at Halsa in Moere og Romsdal County has put into effect new measures in an effort to prevent people from coming too close to Keiko the whale.

> Persons who break the regulations and come into contact with the whale may now be reported to the police. And according to NRK, spectators are respecting the new regulations and keep their distance, and Keiko has started to swim longer distances, which is encouraging, the experts say.

KEIKO COMMUNICATES WITH TYLER GRAHAM

September 7, 2002
Tyler said he still knew very little about what was going on with Keiko in Norway. Again, I initiated the questions for Tyler to ask Keiko.

Tyler: *Keiko, are you there?*
"Yes, I am."

Tyler: *How are you, my friend?*
"I am sad and I am lonely."

Tyler: *Why are you lonely? Where are the people?*
"Because no one is here for me."

Tyler: *They are keeping people away from you because they think it is best for you. We know that is not true, but they think since you've been out swimming and catching fish, you might hurt the people and kids if they swim with you. They also think people will keep you from going back out to the ocean.*
"I would never hurt them. I love them."

Tyler: *We know that you wouldn't, but they don't understand.*
"I know that. It is humans that have helped me throughout my life. This is the least I could do. I want to show them how I feel."

Bonnie: People all over the world have seen pictures of you with the children, and so many, many people understand what you're trying to say. It's not just you and me anymore; there are now thousands of people who know that you want to be with people.
"That's exactly what I wanted."

Bonnie: And, it's happening. It's working! People everywhere are saying that it's obvious he wants to be with people. They're getting it!
"I belong with you. That's why I'm doing this."

Tyler: *They are holding all the people back because they are expecting hundreds of people to visit you this weekend. Even though the people can't get close to you, they are there because they love you and care about you and want to see you.*
"This is my destiny."

Tyler: *Do you feel better knowing that all these people are here to see you?*
"I am being healed as we speak."

Tyler: *Plus, the King of Norway was supposed to have visited you today—that's how important you are!*
"How honored I feel that he visited me."

Tyler: *How do you feel about interacting with people when they come to see you now in the fjord where you are in the water and they're on top—where you have to come out of the water to see them, versus an aquarium where you were in the water and people went down below and could see you through the glass?*
"I would rather be above where the people can see me. I wouldn't mind

doing tricks if it would make the people happy."

Bonnie: So, he would rather see people above the water than through the glass below?
Tyler: Right. That's what I got.

Bonnie: Tell him people can see him when he jumps out of the water, but they're right up next to him when they are separated by the glass. I'm just trying to clarify this is what he wants and that he understands the difference.
"I want to be wherever the people are closest to me."

Bonnie: I truly think he connected more with people when they could see each other up close, than when he jumped out of the water and did tricks. Sometimes he would press his nose on the glass and look right at them. I think there was more heart connection for both. Tell him I'm not trying to persuade him, just trying to be clear. Ask him if he made more of the kinds of connections he wanted to make with people through the glass or when he was outside and people were above the tank?
"I think the people enjoy watching when I do the tricks."

Tyler: *I'm sure they do. We can't promise anything, but which do you prefer?*
"I like to please the people, but for me I'd like to look at the people through the glass."

Tyler: *Another idea is to take you to a fjord that does not freeze in the winter and build a place where people can come to visit you. People could walk out on a pier, then down a flight of stairs bringing them under the water into a room enclosed with glass. That way you could interact with people through the glass and swim free in the fjord. You could swim free if you want to.*
"That's a great idea!"

Tyler: *In the winter you probably won't see many people except those taking care of you?*
"That doesn't matter as long as there are people. That doesn't matter because I love them."

Bonnie: Even if they just build a dock with a six-foot fence on both sides so people could walk out and see him, and Keiko could see them. He could swim around and jump out of the water. It would be easy for people to get to and wouldn't cost that much money.

Tyler: It would be bigger than any park. Do you think he likes doing tricks?

Bonnie: One time Keiko told me, "I would even rather do tricks where people can admire and appreciate my species – thus all whales. It would have a greater influence on how they feel about killing us and ultimately other animals. Our magnificence is hard for anyone to deny."

Tyler: I felt that too.

Bonnie: Once he gave me the idea that people would visit him specifically to practice communicating with him telepathically. So far most have been just looking at an orca whale, and he's a wonderful being, but they would also come to communicate with him.

T for K: Keiko just showed me a clear picture: He was in the water, and people were lined up for miles for wisdom from him.

Tyler: *Keiko, please stay in the water at all times and contact me or Bonnie in a very clear way if you need any help.*

"Thank you. I have no intention of going on land. That was only to get the attention of the people. Now that I have gotten their attention, that is not necessary."

Tyler: *Is there anything else you would like to say?*

"We are from two different worlds—we will come together as one. Our hearts will connect and we will be as one spirit. Thank you so much for letting me talk."

Bonnie: Please thank him, too. Tell him I love him with all my heart and we will talk again soon.

"Thank you my friends."

Bonnie: Tyler, I have to tell you, when I woke up this morning and spotted my stuffed "Keiko" whale among the other stuffed animals. I brought him back to bed with me and held him and touched him. Actually, Susan gave me the idea, so I'm going to hold him and sleep with him every night. You know how much animals and people need to be touched or they cease to thrive?

Tyler: Exactly.

Bonnie: Like people, animals sometimes die when there is no physical or emotional contact, and one of the things Keiko told Teresa last summer was that he misses being touched. He asked her why they don't touch him anymore?

Bonnie: So if you want, something you can do too, is take a pillow or something to bed with you and just imagine you're holding him and touching him.
Tyler: Like, visualize I'm there? I'm touching him.

Bonnie: Yes, you're sending him love and touching him, and he's feeling you.
Tyler: That's wonderful!

Bonnie: I think that'll be very helpful because there's not a whole lot of people in the world that you and I can tell that to. But as you know, he gets it. In fact, in one of our communications Keiko said, "It's nice when people think and care about me. They don't always know it, but I hear them."
Tyler: I'm so honored to be able to do this.

September 8, 2002
The following article by Mark Townsend, September 8, 2002, titled "Keiko's love for children puts his life in danger" is reprinted with permission of *The Observer* (London); copyright Guardian Newspapers Limited 2002:

A dark shape bobbed on the surface of the water as the sun slipped behind the Blue Mountains in Norway last night. Then the object moved, twitching its tail and sending water into the sky. On the shore, men, women and children cheered. Keiko, the world's most famous killer whale, was alive.

But for how long?…

…By yesterday he was less energetic, prompting fears that the latest attempt to reintroduce him to the wild would end in his death. "He is very listless," said Colin Baird, the Icelander in charge, as the mammal lay motionless in the water. "It is all very depressing. He will start wasting away soon."

Keiko has redeveloped a "captive mentality" after people started swimming with him last week. The threat to his welfare forced the Norwegian government to announce an unprecedented emergency ruling yesterday that made it an offence to swim within 25 yards of Keiko or to feed him. Ministers feared a public relations disaster if the health of the world's most famous sea creature declined in waters where it is legal to kill whales."

...Since slipping into this narrow fjord after almost 70 days in the wild, Keiko has again been revelling in human contact, unraveling the biggest attempt to rehabilitate a creature into the wild...

...His apparently successful integration into the open sea went wrong when he saw a fishing vessel off the Norwegian coast and followed it into the Skaalvik Fjord.

Within hours, Keiko had befriended groups of children playing in the water, and within days crowds of them were swimming alongside the killer whale, whose species are seen as fearsome predators in the wild. Some even climbed on his back for free rides.

This triggered fierce criticism from conservationists, who claimed the children's parents were "irresponsible". Mark Berman, assistant director of the Free Willy/Keiko fund, said. "This is a 12,000 lb whale. Would they let their kids ride on the back of a brown bear in Yosemite National Park?"

"Someone could have got hurt. Keiko doesn't know his own strength. These people are making a big mistake." Contact with humans has already made the whale lose its hunting instincts.

The following article is titled "Norway prohibits interaction with Keiko." Jonathan Brinckman, *The Oregonian*, and *The Associated Press* contributed to this report, Sunday, September 8, 2002:

Dave Phillips, director of the Free Willy Keiko Foundation, said the fact that Keiko appears fit nearly six weeks after leaving his pen shows that he has learned to catch wild fish. Now, said Phillips, he and others want the public to stop feeding Keiko so that he continues his transition to life in the wild.

"We think that the best thing for him would be to be in the open ocean with other whales," Phillips said. "We feel that in time he needs to leave this fjord."

Phillips said the foundation had no plans to force him out of the fjord but hoped he would leave on his own if people stop feeding him and interacting with him.

Others are not sure he can succeed in the wild. Stephen Claussen, a former Keiko keeper, said Keiko should have a full health assessment by an independent party to determine whether he is losing weight. If he is losing weight, Claussen said, he should be held in a net pen during the winter and fed.

Claussen said it might not be a good idea to let Keiko return to the open ocean this winter.

Keiko's caretakers began feeding him that morning.

September 9, 2002
Hello Keiko. How are you doing, my friend?
"I am okay. I don't understand what is going on here. I am confused. I can feel the energy of the people, but I could not be with them."

Keiko, your new people believe the public should be kept away so that you will return to the ocean. People have come to see you, and they want very much to see you!
"I still don't understand. Why can't they let me stay with the people?"

Keiko, I will be sending a letter soon to everyone involved to again let them know of your wishes. It seems many responsible for these decisions know too, that you want to be with people. This is very good news.
"Please help me! This is all so frustrating."

Hang in there, Bud. Please be patient and stay in the water.
"I am strong of heart and mind. I can really do marvelous things if people will only notice."

Is there anything else you would like to say?
"I have been waiting for so long for this opportunity. It is all I can think about—my life with people. Thank you for all your help."

Keiko, you are welcome. I love you with all my heart, my friend. I will hold you and touch you and send you my love.
"I am waiting."

Although people were banned from interacting or coming within 165 feet of Keiko, over 4,000 fans, including the King of Norway, made the trip to visit Keiko the first weekend he was in Norway.

On Monday, September 9, 2002 Keiko was taken out for three swims in the fjord. He was also started on antibiotics after blood tests confirmed he had an increase in white blood cells believed due to stress.

The Associated Press, September 9, 2002, reported "Experts feed Keiko the killer whale in Norway as Free Willy star draws crowds":

> Experts have started feeding Keiko the killer whale seven weeks after he was set free in Iceland, while local volunteers help police shield the Free Willy star from thousands of fans along a western Norway fjord, his keepers said Monday.

> They had not fed him since he left Iceland in hopes the black and white orca would tire of human attention and set off to hunt. However, the whale lay listless next to a flimsy dock for several days, raising concerns about his health as he did little more than slowly scratch his back on the bottom of moored boats and occasionally raise his head to greet fans.

Keiko Communicates With Tyler Graham

September 9, 2002
This time when I called Tyler he was in the middle of a communication with Keiko. He read to me what they had communicated so far and then we continued together.

Tyler: *Keiko, how are you?*
"I'm okay."

Tyler: *Is it all right to speak with you?*
"Yes, I need someone to speak with right now."

Tyler: *Are you all right?*
"I'm sad and lonely. I feel as though I am getting sick."

Tyler: *Are you eating enough?*
"That is fine, I just feel sick."

Tyler: *Can you please describe how you feel.*
"I feel there is a hole in my heart."

Tyler: *Keiko, do you think your people know what you want?*
"They think they do, but they are frustrated with me. They are confused."

Tyler: *Bonnie said something is wrong. Could you please tell me what it is?*
"I just don't feel like living without people. I can't go much longer without them."

Tyler: *Keiko, you are such a wonderful, magnificent spirit. I want you to send all of those hidden emotions to me. Let them go.*
Tyler to Bonnie: I'm kind of being his therapist.
"It will take a long time. I have much to release."

Tyler: *Just release it. Let all the feelings go. Express yourself.*
"It will take a long time. I'm just sad that humans and animals don't see eye to eye. It's just that humans think they are superior. They look down on animals. If we just learn to get along and teach each other."

Tyler: *Have you talked to Bonnie a lot lately?*
"I talk to her, but not as much as I'd like to."

Tyler: *I like to hear from you. We all love you.*
"Not everyone, but I love them. All of them."

Tyler: *Keiko you are such a wonderful, incredible example to the people. I just talked with Bonnie. She sends you her love.*
"Tell her thank you so much for her wonderful support."

Tyler: *What do you think the animals should do?*
"There's nothing we can do except be patient and wait for them to understand us. But that is why Bonnie and other Animal Communicators like you can be a voice for all animals to the world."

Tyler: *Do you have any more plans for bringing humans and animals together?*
"Animal Communicators, like Bonnie, understand. They are the biggest part of my plan. The other part is children."

Tyler: *How are the children in your plan?*

"They will lead the people to us and help them understand the animals."

Tyler: *How will the Animal Communicators help you with your plan?*
"By telling the people about the animals. Some WILL NOT listen, but most will."

Bonnie: Please tell Keiko I'm in the process of writing the people involved in deciding what's going to happen with him. Bless his heart, we're in a much better situation because of what he's done, but it's still going to take some time. It's that just his people believe so strongly that he should be free.
Tyler for Keiko: That's been one of his plans.

Tyler to Bonnie: I have gotten it so clear tonight!

Tyler: *Keiko, how are you feeling physically?*
"I'm feeling much better. I'm so happy to talk with you."

Bonnie: Has he seen the people who take care of him lately?
T for K: He has.

Tyler: *How do they interact with you?*
"They are there most of the time."

Bonnie: From what I can tell, they are feeding him now and giving him medicine to help him become healthy again. Is he swimming and moving around more now?
"Much more."

Tyler: *How does that feel?*
"It feels good."

Tyler: *Is there anything else you would like to say?*
"You are my voices to the world. One day we will come together not as humans and animals, but as spirits in joy and eternal friendship."

Tyler: *We will talk to you again soon.*
"I look forward to that."

Bonnie: Tell him we think about him, we care about him, and we love him so, so, so much! We want him to be strong and healthy and safe. And, please stay in the water!
"My heart is full."

September 17, 2002
Keiko, how are you feeling?
"I am better now. I have been wondering what my future holds for me now."

I want you to know that I am trying to let your decision-makers know what you want.
"That's good. I cannot continue as I have been any longer."

Keiko, please understand this will still take time, but so many, many people support you.
"I know. I feel their love again! I have missed their thoughts of me for a long time."

Please describe what your day is like and who you see.
"It is still very lonely. I see people only when they feed me, then all is quiet again."

Do you stay by your people's boat?
"They are not my people. They feed me, but they do not care for me or give me love as my true people did. No, they are not my people.

Do they still take you for swims?
"Yes, we do that often now. Sometimes more than I'd like."

Why is that?
"Sometimes we just go round and round and I receive no reward other than fish.

What other rewards are you missing?
 "The reward from my people. They would talk to me, acknowledge me when I did well and send me their love. It was always fun to do for them. I love them and I miss them."

Keiko, I hold you every night now and touch your body.
"I know you do. I feel your love through my body."

Is there anything else you would like to say?
"Please tell them again and again how I feel until they understand me."

I will always stand up for you, my friend. Please stay safe and healthy and in the water.

"I am always with you in heart and spirit. Thank you for your kindness and all you do."

Keiko's caretakers began looking for another fjord to move Keiko to—specifically one that didn't freeze in the winter. They were also looking for a place where migrating orcas were more likely to visit, and where access for Keiko's fans would be more difficult.

September 22, 2002
Hello Keiko, my friend. How are you?
"I am fine at the moment. I like the interactions I have with people when I can find them. They do still love me. I was not sure there for a while."

Keiko, many, many people still love and care for you.
"Yes, I can feel it again. For a while, I was so lost without their love. No one thought much of me. I can tell they think of me because it fills my heart with love."

Thank you for communicating so clearly. How are you feeling physically?
"Much better. My body is recovering from the ordeal to get here. I just needed some rest to recover. I am better now."

How are you feeling emotionally?
"I am content enough for now. I see people occasionally and would like more, but it's better than before."

Are you eating enough to stay healthy?
"Yes, they are feeding me now so I am getting stronger."

How do you feel about feeding yourself by eating live fish?
"I did what I had to do to be with people. I am not proud of it, but it was what had to be done. Now, I have them feed me."

Do you realize they want you to continue hunting fish on your own?
"Yes, I know that but I will show them I am still Keiko, the captive whale. They cannot change me to be who they want me to be. I am still Keiko."

What happened when they left you with a pod of orcas several weeks ago—when you last saw them in the boat before your long swim to where you are now?
"I was confused. I wasn't sure what direction they went. I looked for them,

but couldn't find them. At first, I was scared, then I saw the opportunity to make my life worthwhile. I thought if I could find a way to see the children and people, I could be so happy again."

Keiko, you are brilliant! Do you know many, many people want that for you too?
"Yes, I know of your efforts to help. I am indeed grateful for your work. We will again see each other some day because of your help."

I hope so much that I will see you again, my friend. Will you be okay this winter?
"As long as they keep feeding me, I will be fine."

Several years ago you said you did not want to leave the aquarium, but you did. Why?
"I trusted my people. I did as they asked because I loved them. These people don't love me and I do not trust them."

Is there anything else you would like to say?
"I am grateful for all those trying to help me find the life I so desire. It is not easy for me or them, but together we can do it."

Thank you Keiko. I love you with all my heart!
"Love to the world."

September 28, 2002
The following article is from *The Associated Press*, September 28, 2002, titled "Norwegian girl serenades Keiko with his theme song":

> Despite the efforts of his keepers, some Norwegians still can't resist Keiko the killer whale, including a little girl who spent hours playing the Free Willy film theme on her harmonica—just like in the movie.

> "That's just the kind of thing we don't need," Colin Baird, Keiko's trainer, said Friday.

> …The friendly orca drew hundreds of fans, allowing them to swim with him, pet him and even climb on his back, until he was so overrun with attention that Norwegian authorities imposed a ban on approaching him.

> Astrid Morken, 8, didn't have to break the rules. Perfectly legally,

she sat on a dock Thursday and played her harmonica until Keiko approached, exactly like the character "Jesse," the little boy who befriended Keiko in the Free Willy movies.

"It was great," Astrid said. She played the refrain from the movie for nearly three straight hours, to Keiko's apparent delight.

…She said Keiko clearly recognized the melody, and almost appeared to dance to it.

…Astrid's innocent passion for the whale illustrates the struggle his keepers face, with everyone from fans to aquariums wanting a piece of the star.

"I'm sorry I can't get all warm and fuzzy about it," Baird said by telephone. He said people had been pretty good about leaving Keiko alone, with local residents in Halsa, the main village of the fjord, even turning out to help keep hordes of fans away from the whale.

But he said when one person can't resist, it sends out the wrong image.

"This isn't about playing with Keiko, or swimming with him, or playing the harmonica for him," Baird said. "It is about trying to return him to the wild…We just want to be left alone."

September 30, 2002

In anticipation of possibly meeting and talking with Keiko's long-time trainer, Jeff Foster, I asked Keiko some questions that might specifically interest Jeff.

Hello Keiko, my friend. Is there anything you would like to say?
"I have been waiting for you again. There is lots for me to tell you."

What would you like to tell me?
"I am enjoying myself more than in a long time. I feel people thinking of me and caring. I love the people so."

Keiko, I may be able to meet soon with your trainer of many years, Jeff Foster. Is there anything you would like to say to him?
"Tell him I am so grateful for his kindness and understanding of me. He truly knows me and our hearts connect. He would always stand up for what was best for me when the others wouldn't listen. I miss him, too. Please thank him for all he has done for me, especially his love and understanding."

Like a magnet, Keiko draws people to him wherever he goes.

Photo: Retna

I don't think Jeff believes it's possible for us to communicate in this way (telepathically). Is there anything you can tell me that might help him believe us?
"Tell him about the time when I was supposed to be nice/friendly to some people and I wasn't. He was very upset with me because it was important to him. I think he has forgiven me now, though."

Is there anything else you would like to say to Jeff and/or the others who were with you for a long time?
"I would like to thank them for all they have done for me. I am indeed grateful, especially for their heart. They really cared about me even though they weren't supposed to. I knew it and they knew it. Tell Jeff I know when he thinks of me, too. I feel his presence and all the good times we had together. Those will always be in my heart along with my memories of him. Tell him I love him the same way even if we are not together."

Is there anything else you would like to say?
"I am much better right now. If only I can stay with people, I would be so happy."

KEIKO COMMUNICATES WITH TYLER GRAHAM

October 1, 2002
Tyler also agreed to ask Keiko some of the same questions to give to Jeff Foster.

Hello Keiko. Are you there?
"Yes. I haven't talked with you in a long time."

Is there anything you would like to say?
"I'm a lot happier since I last talked with you. My cold is completely gone away, and I'm realizing my dream is finally coming true."

What do you mean your "dream is finally coming true?"
"People are starting to understand how animals and humans can come together. That is true happiness and love."

Has anything happened in the last few days?
"Yes, I went to the people. They accepted me. They let me come close to them. They came close to me."

Is there anything about the people in the boat you would like to tell us about?
"They're good people. I don't know if they're afraid of me. They don't touch me very much. They are good people. I could feel their warmth."

Bonnie may be able to meet soon with your trainer of many years, Jeff Foster. Is there anything you would like to say to him?
"Please thank him for helping me through all the hard times in my life, and understanding more about the world. Jeff, what a wonderful man he is. And, opening my eyes to how wonderful you people are. He is a great example to me. He is a great example to us all."

I don't think Jeff believes it's possible for us to communicate in this way (telepathically). Is there anything you can tell me that might help him believe us?
"That's hard to say because if he doesn't, then no one can make him. He has to believe on his own. He can be very stubborn; nonetheless, he's a wonderful man."

Is there anything else you would like to say to Jeff and/or the others who were with you for a long time?

"Thank you—you've helped me achieve such happiness in my life. You've helped me through the hard times. It's like the light at the end of a tunnel. You just have to endure to the end. That's what you taught me."

Is there anything else you would like to say?
"My destiny is with the people. They need me like I need them. The world, we need to pull together. We're all social. We all need love. We all need compassion. We need to learn to be totally selfless. Be safe and may you find happiness."

"Love to the World."

October 6, 2002
I met Jeff Foster at the American Cetacean Society Conference in Seattle, Washington and gave him Tyler's and my communications. I didn't know how he would react to me or my work with Keiko, but he quickly said, "Keiko has made a choice. He wants to be with people." He also told me he was concerned about the limited and inexperienced staff that was caring for Keiko in Norway, and worried how isolating Keiko from people would affect him emotionally.

Even though I was given permission to set out my flyers "What Keiko Wants" detailing my work with Keiko, I was a little apprehensive as to how I would be received at a "whale conference." Would all these people vehemently believe that every whale should be free?

Throughout the weekend, I spoke with dozens of people and, to my surprise, almost everyone said Keiko obviously did not want to be released and that he should be taken care of and permitted to interact with people. I only met two people who felt the attempt to free Keiko should continue: Naomi Rose from the Humane Society of the United States and another woman who was involved with the Keiko Project from the beginning.

I talked one-on-one with Naomi Rose and asked her what she thought about my idea of building a home for Keiko in a fjord that doesn't freeze. Keiko would still be able to interact with people or swim free with wild whales if he wanted. I suggested that people would be happy to pay to see him and, since there are obvious expenses, that would help pay his bills and more importantly give Keiko the life he wants. She responded, "That would be exploitation. We would never do that!"

While on the road to Seattle, I was interviewed on three more televised news programs, and again on Portland's morning television show *AM Northwest*. Through his actions and behavior, Keiko had now verified much of what I had told them he said during previous interviews.

On October 25, 2002 a *CDNN* article reported:

> Norwegian scientists have described Jean-Michel Cousteau's "Free Willy" campaign as a cruel stunt that is doomed to failure and could result in the premature death of Keiko.

> "I absolutely cannot understand that anyone could believe that an animal so influenced by man could become wild again," veterinarian and researcher Finn Berntsen told reporters. "What we are exposing (Keiko) to verges on cruelty to animals."

October 26, 2002
Hello, Keiko. This is Bonnie. Is there anything you would like to say?
"I have missed you again. Time passes quickly here."

How are you feeling physically and emotionally?
"I am fine. I am able to swim and be with my people. I only wish I could interact with more—touch more hearts, open more minds."

Are you eating enough to stay healthy?
"I am doing well that way too. I am hungry at times, but they do feed me."
(I sensed he misses the regular feedings he's had most of his life.)

Keiko, I have new information and ideas for you. Your people in the boat are going to lead you to another place to live for the winter. The area you have been in will freeze over this winter so they are going to take you to a safer place. They will feed you for the winter and you can safely swim in that bay. How do you feel about following them and staying in that bay for the winter?
"If I go with them, will they take care of me?"

They have assured me they will feed you all winter and they will take care of you.
"That is good. I need their help to survive. I cannot do it alone. I especially need their love and attention. Otherwise, I feel so lost and deprived."

Keiko, I'm not sure how much love and attention they will provide, but that is why this communication is so important. I am asking you to look at your situation from another perspective. Because it is winter, it is important that you stay with your people where it is safe. Please be patient and hang in there and take care of yourself. Rest, stay healthy and know that we are closer than ever before to the life you want. How do you feel about that?

"I believe in my heart that I will make a difference in the lives of others; that others will be touched by my love if I am only allowed. I will be there for people who are so inspired."

Keiko, I will help you find that dream, but we still need more time. It is good to dream of what will be. We both need to trust that this is the right path and that it will all work out for you. Right now neither you nor I have total control, so while you are working on it mentally and emotionally, I am working on it in other ways. During this time it is very important that you wait patiently. Have faith and know that it will work out. This winter we need you to cooperate by staying with your people. Eat lots, get strong and stay healthy! Enjoy your new home for the winter, and be ready when the time is right.

"I am humble for your efforts. I will cooperate and do as you ask. I see the bigger picture here too."

Is there anything else you would like to say?

"Yes, when can I look for all of this to happen?"

The soonest will be next spring after the ice melts. If you were to leave the bay to find people, you may become trapped under the ice! That is why it is so important to stay in your new bay, even if your people are gone at times. They said there are fish for you, so please feed yourself to stay healthy. In the spring, after all the ice melts, you can swim to people again! That will be our moment. Do you understand?

"Yes, That makes sense now. I can do that as long as they don't leave me for too long."

Keiko, you can feed yourself in that bay. Please be patient and wait until after the ice melts.

"Thank you again for all you have done for me. I am greatly indebted."

Keiko, I only want you to be safe and healthy and happy, my friend. You have given so much to so many. Is there anything else you would like to say?

"I am enamored and honored by people's response to me. We give to each other in ways not described by words. I will do as you have asked and know that I can trust you. You are my light and hope in this world. Others care, but you are making a difference."

Thank you Keiko. I love you!

With thousands of fans arriving weekly to visit Keiko, in mid-October it was announced that Keiko would be moved from the fjord in the small fishing community of Halsa. He would be taken six miles northeast to Taknes Bay, which does not freeze over in the winter, has more access to migrating whales, and would be more difficult for people to access.

Keiko would be fed at varying amounts at random times and would be allowed to swim to other fjords or into the ocean if he chose.

October 29, 2002
Hello, Keiko. How are you?
"I am doing okay. I miss my communications with you. The mental stimulation here is minimal. Things are quiet again and I miss the people."

Keiko, do you remember our communication the other day?
"Yes, but it is hard for me here."

Please be patient, my friend. More people understand you because of your gallant swim to Norway. You are truly amazing, Keiko! What is important is that you stay in the new bay where they lead you. Please be patient. Together, we will come up with a new plan for this spring.
"I believe my purpose is to influence people and the way they live their lives."

How can you influence the way others live their lives?
"By being in my heart, I show others how to live in their heart. Love is a very powerful force—to be used wisely at all times. Love conquers all. Love is the mystery force of the universe. I hope to teach many others about love. Love of the dolphins, love of the whales, love of all animals and beings of the earth. We are here by design and man needs to realize our gift. We come to man as teachers and healers. We give of ourselves freely and without expectation. That is our nature. That is our purpose."

Thank you, that is beautiful! Keiko, please know that even though they

will feed you, there may not be people with you all the time. Please stay in your new area and wait for them to return!
"I want to do the right thing. Please stay in touch with me. I miss the interaction and need your support."

Keiko, I will try to communicate as often as I can. Is there anything else you would like to say?
"Please be my hero. Show me the way to the life I crave. I am one whale, but together we can influence many people. I am proud and grateful for the many adventures of my life."

Thank you Keiko. I love you with all my heart!
"Thank you the same."

KEIKO COMMUNICATES WITH TYLER GRAHAM

November 3, 2002
How are you my friend?
"Oh, I haven't spoken to you for such a length of time."

Yes, I've been sick.
"I'm sorry. I hope you're feeling better. Don't worry, time will heal all things."

Thank you so much. How are you feeling?
"I'm feeling great. I've had more energy than ever before."

How do you feel about people?
"The people are wonderful. I'm there for them like they're there for me."

Are you still feeling sick?
"No, I'm over that. I'm sorry you are, though."

Does Bonnie speak to you a lot?
"Not very much. I wish she would speak more to me."

She loves you. You know that, right?
"Of course I do. We've been through a lot together."

Like what?
"You know, she helps people understand me."

Do you have any advice for us?

"You are great people and you have so much power. You just need to learn how to use it."

Is there anything else you would like to say to us?
"Stay on the straight and narrow."

What do you mean by that?
"Don't let anything come between you and your goals. Be strong. Be forgiving. Be sensible. I love you all. Message to the world, Keiko."

A Dream

I don't remember exactly when, but one night Keiko came to me in one of the most realistic dreams I have ever had. Actually, it was so real I believe I experienced it at a level that we don't presently understand.

I was swimming in the built-in pool at my childhood home. The rectangular-shaped house paralleled the pool in the back yard. Believe it or not, I looked up to see a killer whale come over the top of the house and land near me in the pool. For a split second, my first reaction was fear—as noted by immediate thought of this being a "killer whale." Not an orca, not even Keiko. Then I realized it was Keiko and my fear disappeared immediately.

Keiko then swam by me, lightly brushing against me. I felt his love. Then he came back around and started to gently pick me up for a little ride. Unfortunately, I woke up. I truly felt he came to me in a way that allowed us to connect on a very deep level. In some way, in some reality, I was with Keiko. I believe that was his way of thanking me, and I could not have received a more wonderful gift.

Of course, the irony of this dream is how the movie *Free Willy* ends with Keiko making a huge jump over a rock barrier into the ocean to freedom. Here, Keiko does the exact opposite, and jumps into a pool (or captivity). Again, I remind you that in general I am not in favor of wild animals in captivity, but I interpreted this experience as what Keiko wanted, and that he was happier in an environment where he was loved, appreciated and could touch the hearts of people.

Keiko Moves to Taknes Bay

November 7, 2002

Unbeknownst to most, on November 7, 2002 Keiko was led by boat to his new home in Taknes Bay. Although the location had been announced, exactly when he would move had been kept secret to avoid more publicity. Locals said this was the most exciting thing that had happened to their remote village in fifty years.

Unaware that this was the day he was moved, I just happened to communicate with Keiko on that day.

Hello, Keiko. How are you, my friend?
"I have missed you too.

Keiko, I am very proud of you. For now, or at least through the winter, you must stay with your people so that you will be safe! It is cold and dark for long periods of time, and the water in many places freezes over so you may be unable to surface for air. Keiko, please, please be patient and stay with your people until we make a plan this spring. Can you do that?

"I will agree to your plan because I know you truly care about me and want the best for me. I can understand these things because I've been with humans for so long. They fascinate me."

How do they fascinate you?
"By being so caught up in their beliefs they can hardly see things for what they are. I am a prime example. People think they know me better than I know myself. If only they would listen to their hearts instead of their minds. Life would be clearer and easier for everyone."

Thank you, Keiko. So are you agreeing to stay with your people until we have a plan for you and it is safe for you to leave?
"Yes, I will cooperate with you."

Do you have any ideas of what we can do to persuade the Free Willy Keiko Foundation and Humane Society people to create a home where you can be with people? (I sensed him being sick) *NO, KEIKO! We want you to always be safe, healthy and happy. We need another brilliant plan like how you swam to Norway to be with people—something like that. Positive, safe for you and that will get lots of media attention. What do you think?*
"I believe in us. I believe we can do this. I want to show the world how animals touch people's hearts; how we can heal each other and make the world a better place together. If man did not kill animals, they would think more about not killing each other."

Keiko, I love you and think of you often! I will keep holding you at night and will communicate with you again through Tyler in a few days. I love you with all my heart, sweet one.
"Thank you for your love and commitment to me. I give my heart to you."

Thank you, Keiko.

KEIKO COMMUNICATES WITH TYLER GRAHAM

November 12, 2002
Again, Tyler communicated with Keiko as I initiated the questions.

Tyler: *Keiko, are you there?*
"Yes, I'm always here for you."

Tyler: *How are you doing?*
"Better."

Tyler: *Can you describe where you are and what you're doing? What's going on in your life right now?*
"Swimming. The people are around me. They see me. They're happy. I'm happy. They're smiling. I'm smiling."

Tyler: *Are you in the same place, or a different place than you have been for the last couple months?*

"I'm in a different place."

Tyler: *What people do you see?*
"Friends that feed me. They accompany me. They enjoy my company just like I enjoy theirs."

Tyler: *Are they there all day and night or just sometimes?*
"Most of the time, only during the day. I'm sad when they leave. Sometimes they stay to watch me. I assume that they understand me on a quicker level. At least I can see them. I can speak and I do receive their thoughts."

Tyler: *Are they on a boat or on land or what?*
"In a boat. They come to visit me. They stay long hours sometimes. I miss them when they leave. There's many different people. I love them all. They are very special."

Bonnie: Does he remember me telling him anything about when they leave?
"Like I said before, I'm not your average whale." (both laugh) "I'm smarter than they think. They think they can lure me away from the people I love so much, but they've got another thing coming."

Bonnie: Tyler, I have asked him, and I'd like you to please ask him again, that when they leave him, that he stay in his new area and to wait for them to return. In other words, I don't want him trying to find his way back to people because as the water freezes this winter he might get trapped under the ice. It is safe there and they will be back to feed him.
"Of course I wouldn't abandon my people, just as they wouldn't abandon me."

Tyler: *Keiko, Bonnie is still working hard on a new plan. Try to be patient and stay there and stay in the water, and wait for your people to come back.*
"I have no intention of going back."

Bonnie: Great, tell him thank you very much and that we love him.
"I know, I can feel the love. I can feel the love in the people."

Tyler: *Is there anything else that we should know about as far as how you're doing or what's going on?*
"I'm with the people and I'm happy to see them. I couldn't ask for anything more."

Tyler: *Are they feeding you enough?*
"Yes, I don't like eating other things, but I need to survive somehow so I eat what they give to me."

Tyler: *How do you feel about eating live fish?*
T for K: I don't know why, but I just got that he doesn't like fish. He doesn't like hurting any other things. A lot of fish are his friends.
"They talk to me, I talk to them. We work things out together."

Tyler to Bonnie: It's like, I wouldn't eat my best friend, but sometimes he has to.

Tyler: *What were you eating before you moved to Iceland?*
"I've been eating fish. I didn't have too much of a problem with it because they're dead."

Tyler: *Do you realize that the fish were alive and then killed for food?*
"Yes, but their spirit has already left their body."

Keiko's perspective reminded me of my childhood farm animals. My family understood they were our pets and that we would never eat them, but we didn't give a second thought to eating the same kinds of animals if they were killed by someone else.

November 13–December 1, 2002
Whenever I communicate with an animal I write the date at the top of the page. Occasionally, when I start, I'm not certain of the exact day so I write the month with the intention of filling in the specific date after I have finished. I failed to date this communication, but it is the next one in my notebook after Tyler's conversation on November 12th, and before my next communication on December 2, 2002.

Hello, Keiko. Is there anything you would like to say?
"I have missed you again. It is lonely here for me. I miss the people and attention from before. I am lonely. There is not much to do."

Keiko, my friend, I am sorry you are lonely and bored. Please know that winter will pass and spring and summer will come. I know how you feel, I don't care much for the short days now either. I keep thinking how spring will come soon and the days will be longer. Please be patient and think of how after all the ice melts, you can again show your people that you want

to be with people. Do you understand?
"Yes, but it is hard for me."

Keiko, millions of people love you and think about you. We will work on a plan for after the ice melts. We need to again show your people in a profound way. Do you have any other ideas for our new plan?
"I feel better when I know we are doing something to change my life. I get very sad here by myself. People have forgotten me again."

Keiko, please know that millions of people do care so much for you. It's just that it's hard for them to help you. They do love you and always remember how you touched their hearts. Are you eating enough to stay healthy?
"I would like to be fed more."

Is there anything else you would like to say?
"I am blessed to be thought of as you do. I wish everyone would understand me as you do. It is hard for me here alone. I will try to do as you ask. Please communicate with me more often now. The days and nights are long and I am lonely."

Keiko, I will communicate with you more often and some with Tyler, too. I pray for you every night and hold you all night long. Together we will find a solution. Please be patient, eat well, stay healthy and where you are, and try to be optimistic.
"Thank you for sticking with me all this time. I am indeed grateful and someday will thank you in person."

Thank you, Keiko. I look forward to that day very much. I will communicate with you soon. I love you, my precious one!

December 2, 2002
Hello, my friend. Is there anything you would like to say?
Bonnie for Keiko: He loves people and wants to be with them! It is cruel to abandon him, keep him from people and the world he knows and loves by insisting he should want to be with other whales. He does not want to be with other whales, and feels rejected and abandoned by the people he knows and loves!

I was also thinking of the outrage both organizations will receive if Keiko dies because they refused to listen to him! (Keiko responded to my thought.)

"You are so right. Those are my feelings exactly. Look at what they have done to me - taken me from the world I know and love only to be alone and isolated. I miss everyone and the joy of being appreciated and loved by others. It is sad. I am sad."

Keiko, please understand that you are closer than ever before to the life you desire, but for the winter it is important that you be patient and stay where you are. Do you understand?
"I do, but it is hard for me. There is so little happening that it is hard for my mind to stay content. I am suffering mentally and hope things change soon."

How are you feeling physically?
"I am okay. It's hard to feel good in my body when I am so sad in my mind. I did so enjoy my time with the people. They show me their love. I have a purpose."

Are you eating enough to stay healthy? (I sensed he really doesn't care.) *Keiko, please eat all they will feed you or at least enough to stay healthy! In a few weeks the days will begin getting longer and soon it will be spring. We will make a plan and you will again show the world. You will again be famous and in the news.*
"It is hard, but I will try, trusting things will get better. I hope people realize the consequences for their actions."

What do you mean?
"I don't think they even realize what they are doing to me and/or putting me through. They believe they are helping me, yet I am suffering. My heart is sad. I only wish they would listen to me."

Keiko, is there anything else you would like to say?
"I am trusting you and those who care for me. I am not able to do much here on my own. Thank you for your thoughts and efforts to help me. I am indeed grateful."

Thank you, Keiko, for all you have given to so many people! Please be patient and stay where you are until we agree on a plan next spring. Always stay safe, healthy and as happy as you can. Know that you are in the thoughts and prayers of many people who wish the best for you. Keiko, I hold you every night and love you with all my heart!

December 10, 2002

Based on Keiko's emotional state during our recent conversations, I was still concerned that he might go looking for people, Wanting to reinforce the message to stay where he was for the winter, I once again solicited the help of Mary and Teresa.

With the long, cold, dark winter days in Norway, I hoped that the more of us who spoke to him, the better chance we had of convincing him to stay where he was. Again, I communicated with him before asking Mary and Teresa for their results.

Hello Keiko. Is there anything you would like to say?
"I am lonely again. I feel like people have forgotten me. I so wish I could enjoy my life with people."

How are you feeling physically?
"I am okay." (I sensed despair; not caring).

How are you feeling emotionally?
"I am sad. My life is pointless now. I don't have much to look forward to."

Are you eating enough to stay healthy?
"Not really. I need more food, but I don't care much either."

How do you feel about your new home?
"It is dark and lonely. I can swim freer but it doesn't make up for the loss of people. I always miss them."

Is there anything you need or would like changed?
"Yes! Take me from here—this world of dark and quiet and take me to the light of people. They inspire me. They give me reasons for living. They are my life. Without them, you have taken away my life."

Keiko, I understand your sorrow. Please be patient and stay at your new home this winter. (I again explained about the ice and sensed he has done that at least once before and it scared him). *Do you understand?*
"Yes, but I'd like to be with people sooner. I don't know how long I can wait."

Keiko, I will continue to communicate with you, and I know others are, too.
"Yes, they are and I am indeed grateful."

Many people are trying to help you to be with humans. Do you have any ideas or suggestions how to convince your people?
"I'm not sure they will ever listen. They know me, yet they do not listen to me. I am tired of struggling with them. They believe in what they are doing so it doesn't seem to matter much what I want. I am supposed to represent all whales, but we are all different. I am Keiko. I am loved as Keiko."

Is there anything else you would like to say?
"Keep on your path. It will happen for us both. I have done all I can from here, the rest is up to you. Best wishes and good luck on your efforts to help me and all animals."

Thank you, Keiko. I love you with all my heart.

KEIKO COMMUNICATES WITH TERESA WAGNER

December 10, 2002
The following is taken directly from our recorded conversation. I stated each question to Teresa, which she then asked Keiko directly:

Teresa: *Is there anything you would like to say? Would you like to talk?*
Teresa for Keiko: Oh yes, he loves to do this almost as much as seeing people. He's getting more accustomed to feeling the love of people long distance. He doesn't like it that way nearly as much, but he is stretching himself to feel that love, so he doesn't feel disconnected from humans.

"It is the way it is. I've come here to be with humans, not simply for my own pleasure or gains—it's what I came here to do. I know many beings are helping me get back to them. Please know that what I continue to say about needing to be with the people is not for my gain. Although I enjoy it so much, it's what I know I came here to do. Thank you all for helping me try to get back to them."

How are you feeling physically and emotionally?
T for K: Physically, he's not as beautiful as he used to be. He's not sick, but he's getting older and he says he is tired sometimes. He doesn't know whether he's tired because his body's really tired or because he does get depressed.

T for K: He says, and he shows me, there's nothing really wrong with his body as far as he knows, he's not sick. The tiredness he speaks of, and the lack of vitality or the lessening of vitality, he really does attribute to just not being as joyful. He has no awareness of any disease or anything going wrong in his body.

And, emotionally?
T for K: His emotions change a lot. In his weak moments he's depressed and sad. He said in his strongest moments he feels great in the knowledge that all that's happening must have a purpose. He's stronger emotionally than he's ever been. He says he will not give up the belief that everything that's happened to him has a purpose. He doesn't believe that he's experienced so much for nothing.

T for K: My impression is that his predominant feeling is of strength. That, that is his base and when he does get depressed or sad, he does come back to the strength. That's the impression I get. He doesn't say that per se, but that's what he shows.

Are you eating enough to stay healthy?
T for K: He says, "I think so." He doesn't particularly enjoy fishing. He really enjoys the ocean. He enjoys the feel of the ocean water all over him—like surrounding him. So it's not like he wants to go back in a tank. He enjoys that part of the ocean, but he really likes being fed.

Are you being fed enough to stay healthy?
T for K: He hesitates and says, "Well, yes." I asked him what he meant because he hesitated. What he showed me is, there are still people who give him food.

Teresa to Bonnie: Now, were you aware of that?
Bonnie: Yes.
T for K: Okay, he's under the impression that they don't give him a whole lot, because he thinks they want him to go on his own and he'd rather not. He'd really rather not.

How do you feel about your new home/environment (where you are living now, compared to where you found the children)?
T for K: Physically it's no big deal; the change. But there are just no people. Physically, it's not a bad change for him, but when he moved he was hoping, "Oh good, the move—People!" And, there aren't any people.

Is there anything you need or would like changed?
"Having my life back again. People, people, people!"

T for K: If he can't have large numbers of people around him, he really wants to see the people he used to know. He thinks some of the people he used to know miss him, and he wants to make them happy. He says, "If you see the people he used to know really well, tell them he loves them."

Bonnie: Is he picking up that they miss him too?
T for K: Yes.

Teresa: *Is there anything else you would like changed?*
T for K: He thinks it's bad to talk about it, because he knows that you're helping him, but he said you know what he wants and needs. He needs people. He thinks he's complaining, and he doesn't want to complain. But he just wants people.

Bonnie: Please tell him I love him.
T for K: He feels your love. He really feels your love. He doesn't take it for granted, but he says he's familiar with your love.

Teresa: *Keiko, please be patient and know that it is important for you to stay at your new home for the winter. There are some areas that freeze in the winter so you may not be able to surface for air.*

Bonnie: I am concerned he might think to go towards people, and I'm assuming as he goes closer to shore there will be more ice.
Teresa: So your concern is that he not get stuck under the ice?

Bonnie: Right. And, this is kind of a conflict because I don't really want to tell him what to do, but he may not be that savvy out there—it's not his natural environment, and he has such a strong desire to be with people. Mary and I came up with just asking him to please stay where he is for the winter.

Bonnie: Also, they are going to be taking him out on ocean swims in January, so we've been telling him "It is okay for you to go on swims with your people as long as you stay with your boat and return with your people." Again, if he gets the bright idea to go find people, this is just not a good time to do that.

Teresa: I think that makes sense. I mean I do that with people's cats and dogs. If someone wants me to tell their outside, free-roaming cat to avoid roads and why, I tell them. So I think there are times, in my opinion, it's okay to be very direct with an animal and say "x-situation" can be dangerous to you. We'd really like you to avoid it. And, that's what you're saying about ice. "If you want to gravitate towards people, we understand that, but if you do that, you can be finding danger with the ice."

Bonnie: Please tell him to be patient and know that there are still lots of people working to help him. Hopefully, he will be more patient and content to hang out where he is for the winter knowing that we are trying to help him.
T for K: He already understands about the ice. He said, "Somebody told him."

Bonnie: I've been telling him and Tyler has been telling him.
T for K: He gets it about the ice. He doesn't want to do that. I told him about being patient. I told him the people will invite him to come with them again in the boat and they might see other whales like him. He said that would be fine. He would look forward to that because then there would be interaction with people. He knows what you mean about being patient. I told him that activity could be about a month away. He said he could do that. He's very aware that you are working for him. He is very aware of that.

T for K: He said in the beginning, it's like he's developed his own telepathic abilities long distance. He just flashed me that when he would be in the tank in Oregon, he would always telepathically connect with children. So it was when they were there, it was like up close and personal telepathy.

T for K: He's telling me that what he's developing now is his long-distance telepathy. He's allowing himself to feel love from you, and feel concern from others. He's letting himself take that in more because he needs it. So he knows you are working on his behalf. He knows there are conflicts; he knows all about them.

Bonnie: Does he have any ideas or suggestions how we can convince his people how to let him have the life he desires?
T for K: First he said, "I don't have any idea." I got the impression he was tired when he said that, so I told him I understand. He said, "I don't understand the politics." I don't understand them enough to say, "try this

or do that." What he did say is just treat everyone respectfully and send everybody love." He said, "I know what happens when you send people love."

T for K: He showed me a picture of sometimes what you and I would call quite an audience where he would have to perform. There would be people that didn't want to be there, or who didn't feel connected to him, or didn't believe he should be in captivity, and they were mad. Not at him of course, but their hearts were closed. What he used to love to do was find the adults whose hearts were closed for any reason and he would send them love and watch their hearts open. To him, it's simple. He doesn't understand the complexity of the politics enough to advise you about what will work, but his opinion is to be respectful and send them love.

Is there anything else you would like to say?
T for K: He wants to tell you, "Thank you." He wants to tell us the earth is absolutely beautiful. And, he knows there's a reason for everything that's happening because he's meant to be here on the earth. This may sound grandiose, Bonnie, but he is so humble. He's very humble. He's just so full of gratitude despite his loneliness. He's showing how beautiful the earth is to him. He's showing people as being kind of primitive, and he says that very respectfully. He shows them as very little and really primitive, but how all we need to do is love them and they will grow up.
Bonnie: You know, I think that's why I'm so dedicated to him. I always felt he just loved everybody. It's like, how can they do this to him? Why not honor him now? All he's ever done is send love.

Bonnie: Teresa, is there anything else you would like to ask him?
T for K: I asked him, *"Is there anything else you would like us to know?"* He said, "Yes, take care of yourself, Bonnie. I don't want to be famous, I just want to love people." He's not opposed to being famous; it's just not his goal. He just wants to love people.

Teresa then asked Keiko a personal question for me.
Bonnie: Back in September when he came to Norway and they took him away from people again, I started sleeping with my stuffed Keiko whale. Now that my cat is very ill, I have been holding him all night. Does Keiko know that?
T for K: He knows about your cat. He's not the least bit hurt that you're

spending time with your cat. What he wants you to do is ask for his help for your cat. He has felt your love and he has felt your touch in that way. And, the fact that you have shifted energy to another being who needs you does not in any way offend him. If anything, he just wants to help. It's really clear who he is and what he is.

Keiko Communicates With Mary Getten

December 12, 2002
Mary e-mailed the results of her communication with Keiko: "Hi Bonnie, Here's what I got from our boy. Hope you're well. Let me know what everyone else gets. Thanks for keeping on top of this. Mary."

Mary: *Is there anything you would like to say?*
"It's good to hear from you."

How are you feeling physically? Emotionally?
"My body feels fine and I am good physically, but I am bored, a little sad and lonely."
Mary for Keiko: I feel that physically he is fine, but emotionally he is a little on the edge. I get that he has hope about the future, but that he will have a hard time alone all winter. His heart is empty and he really needs contact and connection.

Are you eating enough to stay healthy?
"I'm eating enough to stay alive, but I would do better with more food. Could the people feed me?"

How do you feel about your new environment/home (where you are living now)?
"I like this new area; the water is not so rough. I am closer to people here and feel hopeful that people come and interact with me."

Is there anything you need or would like changed?
"More food, more people."

Keiko, please be patient and know that it is important for you to stay at your new home for the winter. There are some areas that freeze in the winter so you may not be able to surface for air. It is okay for you to go on swims with your people as long as you stay with the boat and return with your people.

M for K: Keiko told me that Bonnie had told him about the ice. He said that he can check for that with his echolocation and that he will be careful. He will remember to look for ice if he goes exploring on his own.

There are many people trying to honor your wish to be with people. Do you have any ideas or suggestions how we can convince your people to let you have the life you desire? Please be specific if possible.
"I will continue to seek out contact with people. If I am moved to an isolated area, I will return again and again to the people. Eventually, they will give up. I will find people to be with no matter what! If my keepers want to control my whereabouts, they can provide a suitable place for me to interact with people. If not, I will find my own place. They will see how much I want to be with people."

Is there anything else you would like to say?
"The time has come for humans and animals to know each other on a deep level. We must spend time together. Separation is NOT the way. We are all one, but humans have forgotten this. I am trying to remind them."

February 2003
On more than one occasion Keiko left Taknes Bay by himself. In February, he returned to the fjord near Halsa where he had found children several months before. Keiko's caretakers said they thought he might have "spring fever."

Keiko Gets Trapped Under The Ice

February 27, 2003
Then it happened—Keiko got trapped under the ice!

On February 27, 2003, *The Associated Press* reported "Keiko, a.k.a. 'Free Willy' has ice mishap":

> Keiko the killer whale, made famous by the "Free Willy" movies, learned a valuable lesson last week—not to stray too far under the ice.

> Normally, the six-ton whale lives in Taknes Bay, 370 miles northwest of the capital, Oslo. But last week he strayed to the Skaalvikfjord, 7 miles southwest of Taknes. When Keiko came up to breathe, he was blocked from the surface by the ice and panicked. After several attempts he was able to break through, sustaining some minor injuries.

> "He got some scrapes on his skin, and it will take some time to heal. However, he was really just taking off some skin, and there is no concern at all," Colin Baird, a marine biologist and Keiko's Canadian trainer, told The Associated Press.

> Baird said he thinks Keiko has learned from this experience, and is not likely to get stuck under the ice again.

> "I think he learned a valuable lesson. His appetite is fine, but he is a bit startled. I believe killer whales are able to learn," he said.

March 7, 2003
I sincerely regret that I didn't communicate with Keiko as I had promised. My thirteen-year-old cat was diagnosed with leukemia in the beginning of

December, and I nursed him (often hourly) for the next three months. He and the rest of my animal family also accompanied me to southern California during January and February.

We no sooner returned home than I received news of Keiko getting trapped under the ice. My first thought was, "That's it—this has gone far enough and needs to be stopped before it's too late!" I immediately went to work writing a petition to help draw more concrete support for Keiko.

With the countless people I had spoken with over the last few years who fully supported Keiko being cared for and allowed to be with people, I believed this incident would now surely bring them together in Keiko's defense.

Hello, Keiko. This is Bonnie. Is there anything you would like to say?
"It has been a long time."

Yes, it has. I'm sorry Keiko, but you are always in my heart. How are you feeling physically and emotionally?
"I am drained. I am trying to find people to interact with. I am lonely; just existing."

Keiko, please cheer up. I'm working on a great idea, and very soon—like in the next few days, I'm again going to let many, many people know about you and what you want! It is finally time to ask the people of the world to help you.
"That would be wonderful. I am getting tired of the way things have been."

I'm sorry you got trapped under the ice, and I am very proud of you for saving yourself. Can you tell me what happened?
"I was looking for people, thinking of how happy I'll be when I find them. Then, when I needed to come up for air, I couldn't. I was afraid. I was terrified. I didn't know what to do. I didn't even know the ice would break, but I knew that is where the air is. I got mad. Why am I here? Why are they doing this to me? I just want my life as it was before."

Keiko, I am so proud of you, but you must wait until the ice thaws back into water before you look for people again. Will you please stay in your new home until then?
"Yes, I am scared. I don't have the courage to live in the sea. Please take care of me. I miss my friends, too."

I love you, Keiko. Is there anything else you would like to say?
"Keep trying. Together we can help them understand me. I will wait until the water is safe, then I will go to my people."

Thank you Keiko. I will try to let you know more often how our plan is going. I love you, sweet one, and so do millions of people around the world.
"Thank you for your work to help me."

Keiko, you are very welcome, my friend. Please stay safe and healthy and happy! I love you.

March 13, 2003
I sent Keiko's Petition as a press release to newspapers, radio and television stations, and everyone on my e-mail list.

The press release with a link to Keiko's petition was posted along with all the other articles about Keiko on www.keikonews.com. The petition titled, "Keiko Needs Your Help!" was posted on an Internet petition site through December, 2003.

March 25, 2003
Hello, Keiko. This is Bonnie. How are you, my friend?
"I have been waiting to hear from you. There is lots to tell you. I have been spending time healing and thinking lots about my life situation."

What have you been thinking?
"That I am tired of trying so hard to make them understand me. I can hardly believe that they still won't listen to me. It is frustrating and I am lonely."

How are you feeling physically?
"I am okay. I'd rather eat more fish, but they only feed me limited amounts."

How is your head where you scraped the ice?
"It is healing too. I am not concerned about it. What matters is my happiness. I have not been able to share my love with people and it hurts my heart. I don't understand why all this is being done to me."

They believe they are helping you. Their intentions are good and I have just written them again to try to help them better understand you and that it's possible to communicate with animals. I have also been drawing lots of attention to you with a petition where people sign their support to bring you to a place where you will be safe and with people. You have been getting some attention because of it and it has the potential to give you the life you desire.

"That would be so nice. I miss the people. My heart is sad."

Keiko, please be strong and give us some more time. It's going to take time to organize everyone's support, but it's already started. Everyone loves you and almost everyone wants the same for you as you do. We just need more time. Please be patient my sweet one and stay away from the ice. Once it melts, you can swim to places where there are people. Do you understand?

"Yes. I went too early because I could not stand the lonesome any longer. I will wait."

Keiko, I am trying to find out if the ice has melted yet. Once we know the ice is gone, I will tell you. Then you can swim back to the people and children. Do you understand?

"Yes, I like that idea." (I sensed he was suddenly more positive and cheery.)

For now, you must stay where you have been all winter.
"They take me out with the boat."

That's okay. It may be that they are going to take you out to look for other whales. If so, how do you feel about that?
"I will not stay with them. I don't belong there with them."

Is there anything you would like to say?
"Only that I miss the old days when I was appreciated and people brought me such joy. I am lonely now. The sea and I do not mix. It is not the life I choose."

You must wait until it is safe, then you can return, my friend. I love you, Keiko. Please stay safe and healthy and in the water.
"I will do as you say, my friend. Love to the World."

April 22, 2003
Hello, Keiko. Is there anything you would like to say?
"I am tired. Tired of fighting for what I want and believe in. My heart is sad for the people I miss."

How are you feeling physically?
"I am fine. It is my heart that is sad."

How is the top of your head where you were injured breaking through the ice?
"I have healed well. Not to be concerned with that."

Is there anything else you would like to say?
"I am fine, only my heart is sad. I do hope people will notice me and send their love again."

April 28, 2003
Keiko, how are you doing?
"I am fine. I am exploring/looking for people who will appreciate me."

How are you feeling physically and emotionally?
"Physically, I am fine. I would still like more food. Emotionally, I am better. I am looking for those who love and appreciate me. It makes my life so much more interesting. I have hope when I have people who care."

Keiko, your people may bring you far out in the ocean with the hopes you will stay with a pod of whales.
"Why would they do that?"

Because they want so much for you to be free from humans.
"I know, but they know that is not what I want and it will not work."

If you do not want to live at sea, you probably should turn around and swim back to where you found the people last summer.

Keiko continued to be taken on ocean walks with the hopes of his teaming up with a pod of whales. But for the first time in nine years, neither the herring fish nor the wild orcas that usually follow them, passed by Taknes Bay.

May 5, 2003
Hello, Keiko. This is Bonnie. Is there anything you would like to say, my friend?
"I miss hearing from you. How are you?"

I am okay. I have just started writing a book about you and our

communications. I am so excited about telling everyone your story and helping you to be heard.
"That is wonderful. I am happy for you being so happy. Will you tell them everything?"

Yes, I think people will be so interested in knowing all about you. Is there anything you don't want me to write about?
"Not that I can think of. It has been quite a journey for us both."

How are you feeling physically and emotionally?
"I am better now that people are around me more. I still can't wait for a time that our hearts really touch; when I can see them and feel their energy and love. I had it so good before they took me away."

You know, some people think you must have been miserable because you were not living free in the ocean.
"I know their feelings and in many ways they are right. Those in captivity have chosen to be teachers for the others. With us, it is okay."

What about those in captivity who have communicated they are miserable and want to be set free?
"They must be listened to also. We all are to be listened to and respected. It is a huge lesson for mankind to learn. It is hard for him to respect others."

Why do you think that is?
"Because he has other agendas. Other things he is preoccupied with which seem more important to him. He doesn't have time to watch and see, or feel what is truly in his heart. When he was young, he felt all these things. That is why I love the children so—they are still in their heart. They still feel. They still love."

Thank you, Keiko. That was very beautiful. Do you know that your caretakers are again thinking about leading you far into the ocean with the hopes you will leave with a pod of whales?
"Why do they not listen to me? Why can't they understand that is not what I want? I don't like being alone where no one cares."

They do not believe we have the ability to communicate with each other. They think if you are given enough time, you will realize you would rather be free to swim with other whales.

"Please bring peace between them all so I can live a life of peace. I am tired of this struggle. I don't like conflict. I just want to open my heart to the people who love me. Even though they do not have to see me for this to happen, most think they do. Otherwise, they do not experience it at all. I am happy to be a part of their experience opening their heart."

Thank you, Keiko. Maybe you can keep working on a deeper level to convince your people to stop trying to set you free, and instead allow you to be with people.
"It is a big lesson for them, and they do not yet understand it. It is hard for them to let go of their belief. They are strongly of the belief that they are helping me, yet they refuse to listen to me. Funny how people are sometimes."

Is there anything else you would like to say?
"I want so much to be a good whale. People are so amazed and love me when I am good."

Thank you, Keiko. It was good to talk with you again. I love you, my friend. Please stay safe and healthy.
"Thank you for your encouragement and letting me know what is going on. Please communicate with me more often."

Keiko, I will try. I love you!

KEIKO COMMUNICATES WITH TYLER GRAHAM

May 21, 2003
Again, Tyler kindly asked Keiko questions prompted by me.

Tyler: *Keiko, is there anything you would like to say?*
"I'd like to ask why I haven't heard from you in a long time."

Tyler: *I was busy with school, martial arts and graduating.*
"Oh, I understand. I'm not mad, I just missed you."

Tyler: *Is there anything you would like to say about yourself or what's been going on?*
"I'm just glad to talk to you."

Tyler: *How are you feeling physically and emotionally?*

Tyler for Keiko: There are a lot of fish around. There are people to entertain him and let him watch them.

Tyler: *Are you eating enough to stay healthy?*
"Yes, there is plenty of food over here."

Tyler: *How do you feel about eating live fish?*
"I'd rather eat plants if I could."

Bonnie: Ask him why?
"Because fish are my friends. They are part of one big circle of life."

Bonnie: Tyler, I have a question for you: Do you eat meat or fish?
Tyler: Yes, I love fish.

Bonnie: Do you eat meat?
Tyler: Yes.

Bonnie: The only reason I ask is just to clarify that you're not projecting your perspective on things.
Tyler: No, I like meat. I have never thought of fish as my friend, though.

Tyler: *Keiko, do you know that your caretakers are considering taking you far out into the ocean? They still hope you will meet other whales and swim off to live with them.*
"Why do they do that to me? They should know that I like to be around them; that I wish to be around people."

Bonnie: What does he suggest I do about his situation and to convince his caretakers about the decisions they make for him?
"Let them come to me. Let them speak to me. Let them ask me because they are my friends, but this is my life they are talking about. I wish to be with people. That is my desire; my life. They are my friends and they will understand."

Bonnie: Tell him I don't think they believe it is possible for anyone to communicate with him in this way.
"Why? They must have faith if they want to talk to me."

Bonnie: They probably don't even believe it's a possible concept.
"That's okay. We'll have to find another way."

KEIKO IS KEPT FROM PEOPLE

Summer and Fall 2003
During the summer and fall of 2003, over 200 people a day made the trek by land to visit Keiko at Taknes Bay. And, until a net was put up to keep fans out of his bay, every day dozens more attempted to catch a glimpse of him by boat.

July 12, 2003
Hello, Keiko. This is Bonnie. How are you, my friend?
"I am wandering aimlessly. My life is empty and without meaning or purpose.

What are your days like?
"I travel some with the boat and that I enjoy. The rest is boring and uneventful. I have little interaction with my human friends."

Keiko, I'm afraid it's going to take more time to convince your people what you want—especially since they want so much for you to be free.
"It is their wish. I wish they would understand this."

How are you feeling physically and emotionally?
"I am fine physically. Emotionally, my heart is sad that I am so misunderstood. You would think they, of all people, would know and understand me best. But it is not so. I will keep hoping that they and others will finally understand me and let me live the life I desire with people."

Keiko, you have been the focus of my life for the last five years, my friend. I will only stop trying when you have the life you want.

"I am indeed grateful for your help and the support of others who know in their heart what I am saying."

Thank you Keiko. Please continue to be patient. I will never give up on helping you.
"You are a blessing."

July 23, 2003

On July 23, 2003, Portland, Oregon's *KOIN TV* reported, "Keiko still prefers to swim alone":

> Try as they may, handlers just can't seem to get Keiko to swim with other killer whales.

> …More than $20 million has been spent over the past 10 years to reintroduce Keiko to the wild.

> The project to get Keiko swimming with other orcas costs $500,000 a month at its peak. It has since been scaled down to $500,000 a year. Four on-site keepers track Keiko, and with the help of local fishermen and ferry operators, locate pods of wild killer whales for Keiko.

August 8, 2003

Hello, Keiko. This is Bonnie. How are you doing, my friend?
"I am okay. I miss my close connection with people—when I could really see and interact with them. I was able to send my love directly from my heart to theirs. It was a beautiful experience and time in my life. Now, I know they are here, but it is hard to connect in the same way. I can't look into their heart and touch them in quite the same way. I am honored that so many come to see me. We have touched each others' hearts."

What have you been doing lately?
"I swim in the bay and with the boat. I try to touch the hearts of those who have come to see me. I eat whatever they will feed me. That's about it."

How are you feeling physically and emotionally?
"I like to think I am healthy. My body feels good in the water and swimming. I can hardly wait for the people to come each day. That is the purpose of my life. I enjoy them so. I hear their cheers and sounds as they call to me. I also hear their thoughts and feelings of love—that is the best. I know they love me and keep me in their hearts."

Keiko, what would you think if you stayed where you are and people were allowed to visit you more closely?
"I would like that very much. People make my life complete. I need people in my life—they are my family. I am not interested in living alone in the ocean. My roots are with people. That is the life I have chosen."

Is there anything else you would like to say?
"I have been through this before and I know what will happen. This time I hope to do things differently so the outcome is better."

Keiko, what have you been through before?
"My separation from people. My heart has been broken many times before."

How will you do things differently?
"I will not give up. The power of the people who come to see me give me strength to conquer my sadness. I believe the people will help change the minds of those who decide how my life will be. They are getting closer to accepting my goal/desire to be with people. I am not as discouraged this time."

Keiko, that is very good news! Keep a positive attitude and hopefully they will honor your wishes soon!
"Thank you for your help to make this happen."

You are very welcome. I love you, Keiko and think of you often.
"I know you do. I feel it in my heart."

August 13, 2003
Several people who visited Keiko in Norway during the summer and fall of 2003, submitted letters to Steve Dickey's web site www.keikonews.com. Their observations of Keiko's environment and his emotional state were much like the following which was posted on the "Discussion Forum" August 13, 2003:

> KEIKO: I recently (July 2003), had the pleasure of visiting Norway and was offered the chance to see Keiko in his new habitat. When we arrived...Keiko was in a small cove that had been roped off with two rows of nets. He was floating in shallow water next to a small boat that was tied to a short pier. Access to the pier was roped off so that no one could get near the water or pier. We observed Keiko for about 45 minutes. He barely moved during that time. It was a depressing sight. He looked very lonely.

Later we found a notice put up by a foundation who is feeding him and is supposed to be protecting him from human contact. They are probably the people who put up the nets. We realized that he wasn't depressed because he couldn't leave the cove, but because he was cut off totally from all contact with people and other whales. The notice said that they want him to stop looking for human interaction and they asked visitors not to talk to him or gesture in any way to get him to interact with you. Shortly before we left, he did swim out a way from the pier but promptly returned to the boat next to the pier.

August 20, 2003

There were two interesting articles on August 20, 2003. The first was written by Marc Morano for <u>CNSNews.com</u> titled "'Free Willy' movement was misguided, says former veterinarian":

> …While environmentalists and animal rights activists latched onto the Free Keiko movement and persuaded millions of children to get involved in the feel-good project that began 10 years ago, a veterinarian who once cared for Keiko now believes the attempt to free the whale was misguided.

> …"This is not science; Keiko is a good example of how our actions make ourselves feel good. This is a sad state of affairs," said Dr. Gregory Bossart, a veterinarian, who first had contact with the whale during its stay in a dilapidated aquarium in Mexico City.

> "This whale should never have been released…this is not a release candidate, never was a release candidate," Bossart explained. Bossart is the current director of the division of Marine Mammal Research and Conservation at the Harbor Branch Oceanographic Institution in Ft. Pierce, Fla.

> Bossart's message to the children across the world, who participate in the Free Keiko movement modeled after the fictional whale's life in the *Free Willy* movie, is simple.

> "It's not to believe Hollywood. Hollywood is not what happens in the wild. What happens in the wild is pretty rough and can have ugly consequences," Bossart said.

> The Keiko effort inspired millions of school children to save up money in "penny drives" to help the real-life whale mimic the

fictional whale's adventures from the 1993 movie that culminated in Willy gaining his freedom from an amusement park.

But a decade after the efforts to release him began, Keiko still has not learned how to catch fish on his own and instead relies on a daily diet of about 100 pounds of frozen herring fed to him by human handlers.

…Supporters of the effort to release Keiko discount the criticism and maintain that it's still possible to release the 26-year-old whale into the open ocean.

"We have not lost hope," said Dave Phillips in an interview with CNSNews.com. "I think it's still an open question. He's very strong, he's in very good health," Phillips said.

Phillips pointed to Keiko's 1,000 mile swim from Iceland to Norway last fall as proof of the whale's fitness.

"It still has not been determined whether ultimately he will choose whales or people, but we are continuing to give him the chance to make that choice," he added.

People are kept from approaching Keiko at Taknes Bay, Norway.
Photo: Retna

Another article on August 20, 2003 was by *The Associated Press*. It was titled, "Keiko the 'Free Willy' whale still doesn't want to be free":

> It's been about a year since Keiko was freed from his pen—and swam straight back to human companionship. With the killer whale drawing 200 to 400 fans a day, the bay he calls home seems more like a low-budget "Keikoland" than an experiment in returning a captive orca to the wild.
>
> To keep people from entering the water, Keiko's keepers posted a 24-hour guard and put up orange ropes with "no access" signs along the shore. Temporary nets span the bay to keep small boats out.
>
> …Under the rusted, corrugated roof of a waterfront shack used by Keiko's minders, Keiko T-shirts for sale sway in the wind.
>
> "The perfect thing for us would be to be left alone," says Thorbjorg Valdis Kristjansdottir, a marine biologist who goes by the name "Tobba" and is one of the Hollywood star's four keepers.
>
> But that's not happening, despite the remote, rural location of Taknes Bay.
>
> "There is always somebody trying to get down to the water," says Tobba. "People come at all hours."

September 7, 2003
The Register-Guard and *The Associated Press*, September 7, 2003, contributed to the following article, "Keiko remains under human care in Norway":

> …Keiko's trainers have been hoping that a pod of killer whales would enter Taknes Bay, where the star of three "Free Willy" movies has been living since he showed up unannounced last September. The 33-foot whale might then join them in their migration.
>
> But that hasn't happened, says Mark Berman, assistant director of the Free Willy Keiko Foundation.
>
> "We're still waiting for wild orcas to come around," Berman said. "That should be happening hopefully within the next couple of months, when they start running with the herring again."
>
> Perhaps because of El Niño, the herring didn't swim into Taknes

Bay last season, which meant wild whales didn't follow.

…"They could show up at any time," Berman said. "The public is going to see that dolphins and whales deserve freedom."

By the end of the summer approximately 20,000 people had journeyed to Taknes Bay with the hopes of seeing Keiko. Local merchants reported an upswing in business, and the population increased for the first time in ten years.

Meanwhile, Keiko's ocean walks continued to keep him fit. Wild orcas were expected to swim by Taknes Bay between September and November, and it would be up to Keiko if he wanted to join them. If not, the whales would not return again until the end of February or beginning of March 2004.

October 4, 2003
Another letter posted on www.keikonews.com "Discussion Forum" October 4, 2003, described Keiko's situation:

> I HAVE SEEN KEIKO, in real life! This summer 2003 I and my family went to Norway to take a look on Keiko in Taknes Bay. When we came to the place where he should be, we did not see any killer whale, just a boat in the water. No people at all. But we soon met a dad with two boys. We asked them if they had been seeing Keiko at all. "Yes, he is lying next to the boat," the dad answered. I ran down to the rope they had put up, and looked next to the boat. First I did not see anything, but suddenly he put up his head. It was Keiko!

> When I looked at Keiko I could see that he had banged his head on the ice, but it was going to heal. He did not do so much while we were there. He was just lying next to the boat, who was his friend, and he just swam around a little and sometimes he could put up his head. When I went I turned around a last time, and then he put up his head, like to say good-bye.

> – From a Keiko lover

I was later contacted by BBC World TV correspondent, Malcolm Brabant, who said he had traveled to Taknes Bay during the fall of 2003 to do a feature for BBC World about Keiko's plight. Brabant said:

I shot footage of Keiko moping around the dock and never straying from his handlers' boat.

I was shocked by Keiko's demeanor. I thought he seemed listless and that his spirit had been broken. I observed him on and off over the course of 24 hours and in all of that time he never strayed more than a few meters from the handler's boat waiting for the next hand-out of fish.

It seemed quite clear to me that this whale was never going to return to the wild…and that he appeared dependent upon human company.

While I admired the American effort to encourage him to return to his natural habitat, I thought the time had come for a more pragmatic approach.

I think that the local Norwegians had the right idea...which was to allow him to continue to live in the fjord, but to grant more environmentally friendly access to him.

The fjord was a great home and, in my layman's opinion, a perfect halfway house for a beautiful creature whose mind had been warped by the dreadful seaquaria of the Americas.

October 1–16, 2003

Here is another communication I failed to date. I had written "October" so it was sometime between October 1, 2003 and my next communication on October 17, 2003.

Hello, Keiko. This is Bonnie. Is there anything you would like to say?
"I am so lonely here. There is not much to do. I am too intelligent for such boredom. My mind needs stimulation and interaction with others. This is bad. I am sad."

Your caretakers may take down your net so you can swim out.
"Where will I go? I have no family to go to; no place to go. What will that prove?"

I'm not sure what they expect of you, but if you leave please be very careful to stay away from the ice.
"Yes, I know. But where shall I go?"

I don't know, Keiko. The people in Norway where you are now have been very good to you. They would be happy if you stay. If you swim off, you may get hurt or not be as welcomed. It is too dangerous to go looking for other people this time of the year. Is there some way I can help you through the winter and your boredom?

"Please think of me often. I need reason to believe this way of life will change for me. I need hope that I will be loved and appreciated again. Now, I am lost, lonely and abandoned emotionally. It's like no one cares anymore."

Keiko, I am going to keep writing a book about you. How about if I connect with you each time I start and you can help me?

"That would be great. It will remind me of the times people came to see me and I touched their hearts. I miss the people so much. I am here alone."

Keiko, how are you feeling physically?

"I am okay. I have enjoyed the swims, and I have stagnated more physically than I am now. Even though I don't move much, I have been worse physically."

Hopefully, I will contact you at least every few days. If not, it's because I haven't time to write. Is there anything else you would like to say?

"Be good to yourself. We can help heal each other by deepening our friendship. I will look forward to our times together. It will give me some hope. Hope that things will change."

You know Keiko, one of the main reasons for writing the book is to gather more support from people to let you live the life you desire. Yes, it will be great to have you involved as I write!

"Thank you for remembering me and doing what you can to help me. I will always remember you and what you have done for me."

I love you, Keiko. I will never give up on you, my sweet one.

"I love you, too."

October 17, 2003

The *San Francisco Chronicle*, October 17, 2004, published the following article by Colin Woodard. It was titled "'Willy' trapped between freedom and captivity. Scientists, activists argue whale's fate":

> Keiko the whale is taking a nap, floating like a giant log off the end of a dock on the southern shore of a breathtakingly beautiful Norwegian fjord.

...Keiko swims over to an anchored fishing boat and rubs his back against the vessel for several minutes, causing the 30-foot boat to rock violently at its mooring. Then it's back to sleep.

...Meanwhile, his current providers—the San Francisco-based Free Willy-Keiko Foundation—remain optimistic that he might one day return to the wild.

"We've proved a lot of the skeptics wrong at different occasions," said foundation Director David Phillips, who has been involved with the project from the beginning. "We're not saying that the only successful conclusion is seeing Keiko swim off with native whales, but he's come a long, long way since Mexico."

...For now, Keiko is neither a captive nor truly free, living in a halfway house of sorts in Karsnes fjord, 200 miles northwest of Oslo. To protect him from fans, fishing vessels and other hazards, the gregarious whale spends most of his time in a netted-off section of the bay where handlers feed him 88 pounds of frozen herring daily and take him on regular "walks" around the surrounding fjords with their boat, which he is trained to follow.

...Many project handlers left the project, including operations director Jeff Foster, who believed Keiko preferred human contact to life in the wild. He and other former trainers later wrote letters criticizing Keiko's care, which they said suffered from lack of resources and qualified personnel.

Keiko never came in contact with wild orcas in Norway because they did not come near Taknes Fjord during the 2003 season.

Keiko's Last Communication

November 20, 2003

Little did I know my communication with Keiko on November 20, 2003 would be our last. As you now well know, he pleaded so many times for the life he desired, I didn't realize this time would be any different. Again, I encouraged him to be patient and know that I was still trying to help him.

Hello Keiko. How are you, my friend?
"I am very lonely. Nothing has changed much since we last talked. I wish they would understand me as you and many others do. If they care so much for me why can't they see how sad I am? Why do they do this to me? I am so sad now—so alone."

Keiko, I am so sorry. I have tried so many times to let them know and I am still trying. I won't ever give up on you, my friend. We will get your message out again in a few months. Please hang in there and be patient. Next spring when the book comes out I will make lots of noise for you. In the meantime, please stay safe where you are, so you are away from the ice. How are you feeling physically?
"It's depressing to just lie around all day. Nothing to do; no one to play/interact with."

(I sensed he misses his trainers.) *How are your people viewing your future?*
"They still haven't gotten it. Some are beginning to say/do more to help me, but they are stubborn in their ways/beliefs."

Keiko, I think of you every day and you have totally affected my life since the first day I met you. Next spring we will finally rally the support

needed to give you the life you desire. Please be patient, my friend. Is there anything else you would like to say?
"Keep me in your dreams. I keep hoping they will understand before it is too late."

What do you mean "too late"?
"Before I give up on this new life they want for me."

Keiko, please hang in there. I would be so disappointed if you give up. PLEASE be patient and know that I will never stop trying to bring you the life you desire.
"Thank you. I can be happy here if only people would come to visit. To know I touch the hearts of others makes my life meaningful."

December 12, 2003

On Friday, December 12, 2003 I received a call from Susan Gates. She said, "Bonnie, have you heard about Keiko?" I knew what she was going to say before she told me, "Keiko died today."

Reports were that Keiko followed his keepers' boat to sea on Wednesday, but he appeared lethargic. He stopped eating on Thursday and died Friday afternoon, December 12, 2003.

I was both sad and mad. I had tried so hard to keep this from happening. All Keiko ever did was send love to those around him, and all he wanted was to share his life with people. Keiko was such a beautiful being who didn't deserve to have his life cut short because a few people believed he should live the life they wanted for him.

In some ways I still have not fully accepted Keiko's death. I did everything I could think of to help him be heard, yet was unable to figure out a way to spread his word so that he could have the life he wanted.

Over the years, I made hundreds of phone calls; sent numerous e-mails; wrote letters and press releases; initiated Keiko's Petition; traveled around the states of Oregon and Washington for interviews; and mortgaged my home (twice). In an effort to help him, I even wrote a screenplay believing if people could see and hear Keiko's true story as a feature film, they would certainly rally behind his desire to live his life with people.

Now, as I put this book together, I realize how powerful his words are and that writing this book earlier may have been the way. In hindsight, I also wish I had sent each of Keiko's communications to his organizations as I received them. Although I personally feel I let him down by not doing enough, the claims of success by the organizations responsible for him remind me that there was probably nothing I could have done to change his life.

I know that my purpose in life now is to continue letting others hear Keiko's voice—his thoughts and beliefs, and I will continue honoring him by spreading his message and telling his story for him.

December 13, 2003
News sources from all around the world reported the death of the most famous whale ever. The following articles note some of the events leading up to Keiko's death as well as the controversy about his life:

The Associated Press, December 13, 2003, reported "Movie star whale Keiko dies of pneumonia":

> For kids, Keiko the killer whale was the charming star of "Free Willy." For biologists, he was the focus of fierce debate on whether captive animals could be returned to the wild. Keiko, who died of pneumonia this week, never strayed far from humans, keeping company with them in a Norwegian fjord to the end.

> Keiko's apparent love of human company and his popularity frustrated handlers' dreams that he would one day leave them in search of food on his own. Millions of dollars were spent trying to teach him to survive, but he didn't bond with other whales, apparently feared swimming under ice and died less than two years after he was freed.

> …"It was pretty sudden," his animal care specialist, Dane Richards, told The Associated Press. He said Keiko's handlers went out to check on him during a late afternoon blizzard and he was still alive. Two hours later, he had died.

> …David Philips, executive director of the San Francisco-based Free Willy-Keiko Foundation, said Keiko's plight changed public perception of whether a whale could be returned to the wild.

> "We took the hardest candidate and took him from near death in

Mexico to swimming with wild whales in Norway," he said. "Keiko proved a lot of naysayers wrong and that this can work and that is a very powerful thing."

But after 25 years in captivity, Keiko appeared to prefer human companionship…

…The popularity made training a struggle for his keepers, who had been trying to keep fans away in the hope that Keiko, feeling a need to socialize, would seek out wild killer whales.

But people still came to see him, and Keiko seemed to like it.

"He was like the family dog, he wanted to be next to you," said Mark Collson, a board member for the Oregon Coast Aquarium.

…Nick Braden, a spokesman of the Humane Society of the United States, said veterinarians gave Keiko antibiotics after he showed signs of lethargy, but it wasn't apparent how sick he was.

"They really do die quickly and there was nothing we could do," he said. "It's a really sad moment for us, but we do believe we gave him a chance to be in the wild."

December 14, 2003
The following article, "Activists debate returning captive animals to the wild," December 14, 2003, is reprinted with permission of the *Statesman Journal* and *The Associated Press.*

…"Keiko showed us that it's possible to return an orca to the wild," said Naomi Rose, an orca biologist with the Humane Society of the United States, which co-managed the project with the Free Willy-Keiko Foundation. "He didn't swim off into the sunset, but for the last five years he thrived in his natural environment."

…Orcas live an average of 35 years in the wild, and it wasn't clear how much Keiko's time in captivity—or his reintroduction to the outside world—contributed to his death.

"It's a foolish game to try and predict or figure that out," Rose said. "Who knows? I firmly believe he lived as long as he did because of the care he received. He had a pretty rough life in captivity."

"In the wake of a whale," by Katy Muldoon, *The Oregonian*, Sunday, December 14, 2003, Keiko's former trainer reminisces about Keiko:

Stephen Claussen zipped on a wet suit, strapped on a scuba tank and jumped in. The Newport night was so clear, he remembers, that from underwater in Keiko's Oregon Coast Aquarium tank, Claussen could see the stars. The killer whale joined him near the pool floor, and for 90 minutes the 21-foot-long beast let Claussen scratch and rub him, stern to stem.

"It was magical," said Claussen, who worked with Keiko for six years in Oregon and Iceland, "because he was such a gentle creature and he genuinely seemed to enjoy the interactions with people."

That magical quality and ability to endear himself to humans ignited the most remarkable, quirky and controversial animal rescue in history—an effort that ended Friday, when Keiko died in Taknes Bay, Norway. He was about 26 years old. The cause of death is unknown, but his veterinarian said that symptoms indicate Keiko may have had acute pneumonia.

…"Keiko was an excellent ambassador, not only for his species for but all the creatures in the marine environment," Claussen said. "He really brought human awareness to a new level."

…Visitors young and old streamed by the underwater windows that provided a good look into his tank. Many, mesmerized by the sight of him, would stand nose-to-the-glass for hours. Often, Keiko would stare back.

"We humans have an incredible connection to dolphins and orca whales," Craig McCaw, the project's chief benefactor, said at the time. "Look into their eyes, and you know that they are intelligent and special...Our moral imperative was to do our best to achieve as much freedom for him as he wanted or was capable of after all these years in captivity."

KEIKO'S SECRET BURIAL

December 15, 2003
During a heavy snowstorm in the dark of night, Keiko's burial began. An excavator dug a large hole, and then with the help of two tractors took Keiko from the waters that had been his home for the last year.

Because of freezing temperatures, Keiko easily slid across the ice into his grave, which was quickly covered by snow. Seven people, including Keiko's caretakers were present during the burial, which ended about 1:30 Monday morning.

Keiko's organizations said the burial was done in secret to avoid a "media circus."

December 16, 2003
The following article, "HBOI scientist says 'Willy' star should not have been freed," December 16, 2003, is reprinted by permission of *Scripps Treasure Coast Newspapers*:

Harbor Branch Oceanographic Institution's Greg Bossart on Monday remembered the first time he met Keiko, the killer whale featured in the movie "Free Willy," who died last week.

Bossart had been called to a Mexico City aquarium by another veterinarian to check out the orca's large skin lesions—which he noticed when Keiko starred in the 1993 movie.

"I knew this whale was special when I first met him," said Bossart, director of Harbor Branch's marine mammal research and conservation department. "I remember I put my legs in the water, and Keiko put his head in my lap, like a big Labrador retriever."

Knowing how comfortable the killer whale was with humans, Bossart said he publicly opposed the push to free the 6,000-pound Keiko a decade ago.

Now, he said, Keiko's death from a quick onset of pneumonia off Norway on Friday proved that freedom is not always a happy ending for every wild animal.

"This was a failed and flawed experiment from the beginning," he said in a telephone interview. "As far as I'm concerned, Keiko was never a release candidate. He was caught in the middle of a Hollywood movie."

In the early 1990's, Bossart, who was a consultant with the University of Miami's pathology department, got to know the killer whale studying his lesions for nearly three years in Mexico City.

With that work, he was able to identify the first case of papilloma virus in killer whales, and helped author a published paper on his findings.

After the whale was transferred from the small pool in Mexico City to a more accommodating one in Oregon, Bossart was quick to speak out against school fundraisers and efforts by other scientists to release Keiko back into the wild after 25 years in captivity.

"I was one of the few public scientists to say he was non-releasable," he said. "I think every little child in America hated me. But we as human beings need to understand what our actions mean for the animal," he said.

Keiko was released into the wild from Iceland in July 2002, and the whale swam straight for Norway, where he allowed people to pet and play with him.

He died at age 26. Orcas normally live an average of 35 years in the wild.

Some wild animals can be returned to the wild without any problems, but scientists have to be sure that the animals are healthy enough and able to fend for themselves, Bossart said.

"The fact that (Keiko) spent all this time not being able to feed himself, that he never integrated back into the whale society and always sought the attention of humans proves this was a flawed experiment," he said.

"Keiko was caught in the middle of it, and ultimately he died for it," Bossart said.

An article in the *Newport News Times*, December 16, 2003 titled "Impact of Keiko's life felt worldwide" by Steve Card, features an interview with another trainer, Peter Noah, who worked with Keiko at the Oregon Coast Aquarium:

…Had Keiko remained at the aquarium in Newport, Noah believes he would have died sooner."

Noah believes, however, that one flaw surfaced in the Keiko project. "It was always banked upon that he would swim to another group of whales and join up and leave. They never looked at the other side of that coin," he said.

Public statements on many occasions noted that the choice of swimming with wild whales or remaining with his human companions was a decision that would be left up to Keiko.

...But those in charge of his care in Iceland, and later in Norway, were focused only on his release, said Noah, even when it became more and more clear that Keiko was choosing people over whales.

"What happened is that the organizations were never, ever really committed to that second contingency (staying with humans)," said Noah. "What happened is that Keiko essentially made his choice, and they didn't respect it."

"The aquarium, the Free Willy-Keiko Foundation, the humane society, they never really looked at Keiko as Keiko," continued Noah. "They looked at him as a cause, they looked at him as an agenda, they looked at him as all those things that could get their organization further along. That's the sad part of it."

* * *

KIRO 7 Eyewitness News aired the documentary "KEIKO: The Inside Story" on April 24, 2004. It was written and produced by Penny LeGate who interviewed the key people directly involved with the Keiko Project, and discussed how far the effort to free Keiko should have gone.

The following are excerpts from the documentary with Penny LeGate interviewing Keiko's long-time trainers, Jeff Foster and Stephen Claussen; Naomi Rose, marine mammal biologist with the Humane Society of the United States; Dave Phillips, director of the Free Willy Keiko Foundation; and major financial contributor, Craig McCaw.

Discussing Keiko's behavior in Iceland:

Penny LeGate: Time and again, Keiko was brought out to mix with wild orcas.

Jeff Foster: We were kind of banking on the wild population, and these animals being so social that hopefully he would learn from those animals, but he never did seem to pick up on that.

PL: The staff said in these cases sometimes the killer whale pods would swim directly towards Keiko, but then sharply veer away. Keiko would vocalize aggressively and scare them off.

Stephen Claussen: When we left him one time, we removed the walk boat and let him just be out on his own. We would go out in the helicopter and track him and find him. I would see him hanging out with fishing buoys—and there would be whales all over in the area and, no interest.

JF: Yeah, I think he had been in a captive situation, and he had been dependent on people so long that this was foreign for him. We don't know until you try something like this.

PL: After two years of trying, Keiko does not swim away with wild whales.

SC: We also seemed to see the writing on the wall that after two full seasons of going out again and again and again, with hundreds of interactions, that he was making a conscious decision that either he could not or did not want to reintegrate with these other animals.

PL: In the summers of 2000 and 2001, Keiko's trainers led him out to whales dozens of times. The encounters are always brief—lasting ten to twenty seconds.

JF: Either the whales would take off going one direction, or Keiko would take off.

SC: After three years, he seemed to have made a very clear, conscious decision: "I want to be with people."

PL: From the very beginning, the promise was to take Keiko only as far as he wanted to go; that every step would be his choice. At this point though, some believe that certain agendas are forcing those choices.

JF: I think that in certain circles, you know, "release at all costs" was felt like that was very important. That was the goal—to release this animal: "Better off dead than fed."

SC: People got so overwhelmed by emotion that it went from the scientific project to see if this could done to: "He has to be free at all costs." The social interaction that Keiko knew and obviously, in my opinion, had chosen, was people for the interaction. And, he should have been allowed that.

Naomi Rose: We just wanted to show that putting a long-term captive killer whale, orca, back into the wild was not a death sentence.

PL: Did he have a shot at all in Norway?

JF: No. Not even a little one. We knew that he spoke Icelandic dialect with whales, and that when he went to Norway he wouldn't understand that dialect.

NR: Did I really believe he would fully integrate? No. Did I think he'd probably end up being sort of in a semi-wild, semi-captive state for the rest of his life? Yeah, I was reaching that point. Were we getting to the point where we thought well, this is about what it's going to be like for the rest of his life? Yes.

Craig McCaw: He closed a chapter and you couldn't help but be sad. But at the same time, happy that he died free and in a way so much happier than he started.

PL: Yet some feel that the continued push to set him free deprived Keiko of the human contact he sought time and again.

SC: I felt in all likelihood that he died a lonely death.

JF: I like not to think about that, but I think there's that possibility. I know the staff interacted with him some, but a minimal amount because they wanted to try to keep him from focusing on the people above water.

Dave Philips: I don't believe Keiko was lonely when he died. I don't think he was in any less contact with people than he had been in Iceland. He was in a place where he could come and go. He was free.

JF: He was a marvelous animal with a wonderful personality—always willing to please. He was a very, very special, special whale.

On April 16, 2004, I sent both the Free Willy Keiko Foundation and the Humane Society of the United States a copy of Keiko's Petition along with the 650 signatures from people around the world who wanted Keiko to be allowed to interact with people. I did not receive a response.

KEIKO'S MEMORIAL & SYMPATHIES

December 2003

Sympathies came in many forms during the rest of December. I sent an e-mail to everyone who had previously shown an interest in Keiko. Below are a few excerpts from my "Last Request for Keiko":

> I would like to thank each and every one of you for your support of my work and your love for Keiko. I'd especially like to thank the many people who made a special effort in one way or another to help me try to help Keiko.

> The following is from a sympathy card I received from a friend— I'd like to share it with those who loved Keiko:

With Deepest Sympathy
IN THE LOSS OF YOUR ~~FATHER~~ *WHALE*

(Inside)

May you always feel your
whale ~~father~~ is with you...
for he lives in your life
and the lives of
of all those
who knew and loved him.
Sincere Condolences

(Signed)
Friends of Keiko

I am the Animal Communicator who since 1998 has tried to let people, and the organizations responsible for Keiko, know that he did not want to be set free. I have devoted the last five years trying to avoid exactly what has happened!

I believe if Keiko's simple request had been honored, he would be alive today. All he asked for in his communications and through his obvious behavior was to be with people. Imagine how happy and content he would have been if people had been visiting and interacting with him. Many people have told me Keiko died of a broken heart. I agree.

I fully believe the initial intentions of the Free Willy Keiko Foundation and Humane Society of the United States were honorable, but when it became obvious that Keiko did not want to be free, it was time to admit they had tried what they thought was best, and then to honor what Keiko wanted. By the sounds of their comments, if they had it to do over again, they would do the exact same thing. Please learn from Keiko's sacrifice, so such a tragedy does not happen to another animal.

In response, I received many e-mails regarding Keiko's death. Below are a few examples:

Hello Bonnie,

I received your recent e-mail and request to send letters to Dave Phillips and Naomi Rose to try to get them to listen to and become more open-minded about our ability to communicate with animals.

As I continued to read your e-mail and your response to the comments that these people made to justify their actions as well as proclaiming to the world what a successful venture it was: In THEIR minds it was a success because they accomplished what they set out to do, which again in THEIR minds was the right thing. The fact that they are narrow-minded is another issue and the fact that they do listen to others who also have the well-being of the animal in mind is also an issue. Most people who have any kind of power over another person or animal appear to have the inability and respect to listen to others.

I see it all the time with parents who are trying to force their child to be and do what they want and, because they are the authority, the child is almost powerless. When it comes to animals, they are powerless because of their inability to defend themselves or speak up for themselves. The greatest lesson for all mankind is the ability to respect each other, whatever our beliefs or values, and to remember that we are only guests on this planet and do not OWN the world or our children or the animals.

Asking people to be respectful to each other, the earth and the animals that share this planet with us is the biggest challenge of all and I will give a lot of thought as to how I can articulate this even further and will be more than happy to write to these people to let them know how they enforced THEIR will without respect for the wishes of those who also cared for and loved Keiko, irrespective of whether they believe in animal communication or not.

The fact that you devoted so many years of your life to help Keiko and are passionate about helping to expand the consciousness of mankind is to be honored and appreciated. Keiko's loss would be been in vain had you not invested so much love and labor to help him even though it fell on deaf ears!

Zdenka

* * *

Dear Bonnie:

I can't even come up with words to tell you how sorry and angry I am that Keiko had to die without ever realizing his dream of living with people. What a sad world it is that humans couldn't love him back enough to allow him that dream.

You did so much to help. Thank you.

Many regards,

Cyndi

* * *

Merry Christmas Bonnie ~

What a sad thing it is that those who are in charge of the welfare of an animal do not seem to listen to what that animal wants. Even if they do not believe in animal communication, it should have been obvious by his body language that Keiko was not happy.

Speaking as an Oregonian who was quite upset when Keiko was taken from the aquarium he enjoyed so much here, it was very sad to hear of his death. I have no doubts that he was much happier when he was in a place where people could interact with him and enjoy his lust for life. You would think that an organization like the Humane Society would want to have people on their staff that care enough to learn to communicate with the animals and ask what they want.

For years scientists have been searching for measurable intelligence in animals and trying to learn how they communicate. Unfortunately, they seem to be unable to accept the idea that animals and humans can communicate on a level that cannot be validated by science at this time. I am not an animal communicator —I do not have any training or experience in this area other than my interaction with those who have learned this skill. Still, I am willing to accept the possibility and compare what I am being told the animals are saying with what I observe personally in them and their behavior. Keiko is a prime example of this.

Even if the message received was not exactly right, the core of the message was obviously correct. I hope that in the near future, the attitude of those who have been given the all-important duty of caring for these animals will change. It is not what you think is best for the animal that is important; it is what the animal wants and needs to be happy. Our society is so willing to believe in life after death and communication with those loved ones who have passed on, as evidenced by the popularity of shows like "Crossing Over" with John Edwards. It is a shame that this belief cannot be transferred to the animal kingdom, especially those animals who show such remarkable intelligence like dolphins, whales and chimpanzees, just to name a few.

I hope that what I have said here is taken to heart by organizations like HSUS, who are supposed to care about what is best for the animals. When you do what you think is best and it is obviously not working, maybe it is time to try something else, like listening to what the animal wants. Open your mind to the possibility and maybe you can avoid tragedies like this in the future.

Peace and love,

Doug

During the month of December 2003, www.keikonews.com "Discussion Forum" also received a number of comments from those who supported the effort to free Keiko and those who did not:

(Date: 13 Dec. 2003) I'm sorry but I disagree that the "right thing" was done for Keiko. He was a domesticated animal and the reintroduction process did not go as thought. He went in search of more human companionship. He must have felt abandoned—living out his days alone—keeping the people who loved him away from him. He never really adjusted to the "wild"—he always had to be fed. I dare say he would be alive today had he been left at the Oregon Coast Aquarium where my husband and children went to see him—a 2,000 mile drive for us.

Well, I guess he can't be hurt anymore now. I have just cried my eyes out thinking of how lonely he must have been—maybe you think I'm crazy and don't know how he felt—but just MAYBE ALL THE DO-GOODERS DIDN'T EITHER. He was DEPENDENT on humans. I know you all fed him and tended to his health—but gosh darn it! He was a SOCIAL being and I feel he really missed that part of his life. Twenty years is a long time to be treated one way and suddenly be expected to behave like a wild animal.

Sorry, but I am just rather bitter about the whole thing. Why should a whale be treated any differently than a tiger or lion which was kept in captivity? They are NOT returned to the wild. They are kept in animal parks where they are cared for the rest of their days. Keiko deserved at least as much. I'm glad my family had the privilege of seeing him in Oregon, where, I believe, he should have remained. God bless Keiko.

* * *

(Date: 17 Dec 2003) Why didn't someone listen to Greg Bossart of the Harbor Branch Oceanographic Institution. He knew more about Keiko and what was best for him in 10 minutes and he is right—Keiko was caught in a Hollywood movie and everyone idealized what would be best for him. Poor Keiko—yes, he was free—free to be alone. He so loved people visiting him and talking to him. I agree we should not capture any more whales—but PLEASE—don't ever do this again. I think Keiko was just too old to do this. He should have stayed in Newport, Oregon. He was doing well there and he would probably still be alive.

It makes my heart hurt. At least in death you finally respected his feelings and didn't push him out to the 'wild' to be devoured by other beasts of the sea. He was just the most special animal ever. We loved him. God bless him. If there is a place in heaven for whales—he is there, no doubt about it.

* * *

(Date: 28 Dec 2003) I was so sad to see that Keiko died. I wish we would have left him at the Newport Aquarium. He would still be alive. He was so used to human companionship. We all loved and supported him in Oregon. I know that sometimes we think we are doing the right thing, but it turns out to be wrong. Sad, sad, sad.

January 2004
In January, about three hundred Norwegian schoolchildren went to Keiko's burial spot at Taknes Bay. They said their good-byes and placed stones on his grave, creating a burial mound.

Keiko's Memorial Cermony

February 20, 2004
On Friday, February 20, 2004 the Oregon Coast Aquarium held a candlelight memorial service for Keiko. Over six hundred people crowded into the facility to honor one of the most precious beings to bless our planet.

An article by KATU.com and *The Associated Press*, February 20, 2004, described the evening, "Ceremony, display, bring closure for Keiko's fans":

Keiko the killer whale, star of the Free Willy films of the 1990's, was memorialized Friday night as one who bonded with humans and taught them patience and trust.

Candles were lit, there were songs and poems and film of Keiko swimming free, shown in public for the first time.

…Veterinarian Dr. Steve Brown, who had worked with Keiko in Newport and Iceland, said people bonded with Keiko in a way he had never seen.

"And Keiko's bond with people was obvious," he said. "The bond went both ways."

Thomas Chatterton, a chaplain who often deals with veterinary issues, said Keiko "was not one of our kind, but nevertheless he was still one of us."

He asked why people should care so much for a whale when there is poverty and homelessness in the world. "He was a friend," Chatterton said. "He endeared himself to millions of people."

Over 600 people crowded into the Oregon Coast Aquarium to attend a memorial ceremony honoring Keiko.   AP/World Wide Photos

The following article, "Hundreds of fans join to bid Keiko farewell," February 21, 2004, is reprinted with permission of the *Statesman Journal* :

> …Mothers and fathers hugged their children and wiped tears from their eyes as trainers and veterinarians, voices cracking with emotion, talked about their work with Keiko. A photo montage showed Keiko with his signature floppy dorsal fin and three freckles jumping out of the water, rolling on his back and acting playful, just the way many remember him in Oregon.

In many ways the memorial service was typical—with prayers, hymns and readings. That so many came to pay their respects to a different species made it one-of-a-kind.

…After the service, people looked at a display of photos of Keiko and letters for him. They viewed oil paintings of him. They wrote their final farewells in a book that couldn't possibly hold everyone's thoughts.

"Such fun it was while you were here," wrote an anonymous Keiko fan. "Such sadness in your being gone. May your spirit live on."

"He wasn't just a whale," said Salem resident Travis Dudley, who attended the service with his family. "He was almost like part of the family. As you can see from all the tears, he wasn't just a whale. Just look at that face. Who couldn't love him?"

Keiko's Message to the World

August 16, 1997 & March 1998 - at the Oregon Coast Aquarium

"We would like man to realize what he is doing to the world. That many changes have taken place over years and there are more things to come. Man must realize that he has a tremendous effect on the future of our planet."

"We would like man to understand that we all want to live in peace—to be free from fear and pressure to survive. We want all animals and man to love alike. There will always be differences but those can be worked through in a peaceful way—without violence and war. Understanding and compassion are the answer to peace. If man only knew how much he is influencing everything on this earth, and that everything he does will leave an impact on all other living beings."

"Many will listen and learn and grow from what they hear from us. We are all here with a/our specific purpose to learn and grow and be who we truly are. To learn to love and forgive and enjoy our life on this earth. People are stumbling through their life with no awareness of their purpose or the big picture."

"Learn to live in peace and take care of our world. It does not exist just for man alone. Many others depend on the earth and its survival for their existence. Man believes he is the special one and all others are less significant. Without the rest of us, man is nothing – he will cease to exist."

"I am quite content for the moment. I am feeling much better and healthier than before. I like the attention and effect I have on all the people that come to see me, especially the younger ones and the children. They will remember seeing me for a long time."

"My goal is to let the people of the world learn by my captivity. Learn what not to do to us. Learn that we are born free to stay free—never to be caught and made fools of. We are far more than man realizes."

"If they set me free because they believe they are helping me, they will be doing a disservice to many, many people that will be influenced. I need to impact others by my captivity much more than I need to be set free. If they insist on setting me free, I will continue to do things and/or behave in a way to discourage them from letting me go for my purpose is best accomplished by being here."

"Please let them know this. They believe they are doing what I want, but it is what they want for me. They want me to be something I am not. I am not a wild whale; I am Keiko."

"Love to the World from Keiko."

* * *

www.animalmessenger.com

www.keikospeaks.com

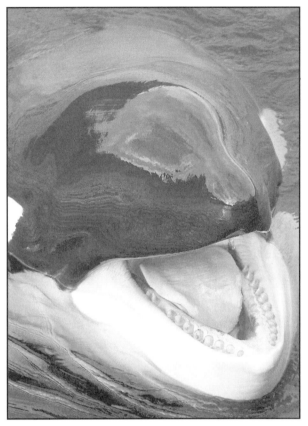

Keiko: "I am not a wild whale; I am Keiko."

AP/World Wide Photos

* * *

I have written this book with the hope that Keiko's death will not be in vain because people will be touched by his message and become aware that it is possible for us to communicate with animals. By learning to listen to the animals and honor them as individuals, we thereby enrich their lives and ours.

* * *

"The greatness of a nation and its moral progress

can be judged by the way its animals are treated."

~ Mahatma Gandhi